OFFSHORE RENEWABLE ENERGY

Wave, tidal and offshore wind technologies have long held the promise of seemingly limitless energy supplies. In practice, while offshore wind is growing relatively rapidly, all three sectors have lagged behind expectations. This book, from the International Energy Agency's Implementing Agreement on Renewable Energy Technology Development (IEA–RETD), examines the reasons for this and suggests how barriers to deployment might be overcome.

Beginning with an assessment of the marine energy resource, it provides a detailed introduction to the main technologies currently being employed to harness wave, tidal and offshore wind power. It then examines the types of policies which are used to encourage deployment around the world, and progress towards meeting targets. The economics of offshore energy projects are discussed, along with risks that projects face and the types of finance which are available. A final chapter turns to barriers – both technical and non-technical (including environmental, health and safety, skill related, supply chain and more) – and in all cases suggests how to mitigate and remove these barriers.

Highly illustrated in full colour, this is an indispensable resource for anyone – whether in industry, policy or academia – looking to learn more about how deployment of offshore renewable energy technologies can be encouraged.

The IEA–RETD is one of the key outcomes from the International Conference for Renewable Energies in Germany in June 2004. Members of the IEA–RETD are countries that want to encourage the international deployment of renewable energy through improved policies. Complementing the other International Energy Agency's implementing agreements on renewable energy technologies that focus on specific technology research, development and demonstration issues, the IEA–RETD is a technology cross-cutting and policy-focused platform to accelerate the deployment of renewables (see more at http://www.iea-retd.org/).

OFFSHORE RENEWABLE ENERGY

Accelerating the Deployment of Offshore Wind, Tidal and Wave Technologies

The IEA-RETD has appointed Mott MacDonald to carry out this project. Mott MacDonald is an international, multidisciplinary management, engineering and development consultancy serving the public and private sector around the world.

Report authors: Kolliatsas, Christos; Dudziak, Gregory; Schaefer, Julia; Myers, Noah

RETD
IEA - RENEWABLE ENERGY TECHNOLOGY DEPLOYMENT

Mott MacDonald

earthscan
publishing for a sustainable future

First published 2012
by Earthscan
2 Park Square, Milton Park, Abingdon, Oxon OX14 4RN

Simultaneously published in the US and Canada
by Earthscan
711 Third Avenue, New York, NY 10017

Earthscan is an imprint of the Taylor & Francis Group, an informa business

The content of this book uses data and information valid at the time of writing (December 2010)

British Library Cataloguing in Publication Data
A catalogue record for this book is available from the British Library

Library of Congress Cataloging in Publication Data
Offshore renewable energy : accelerating the deployment of offshore wind, tidal, and wave technologies / IEA-RETD. -- 1st ed.
p. cm.
Research in support of the International Energy Agency's Renewable Energy Technology Development (RETD) Implementing Agreement.
Includes bibliographical references and index.
TJ163.2.O39 2012
333.91'45--dc23
2011028392

ISBN: 978-1-84971-470-9 (hbk)
ISBN: 978-0-203-13884-7 (ebk)

Typeset in 10pt Frutiger
by Saxon Graphics Ltd, Derby

Printed in Great Britain
by Bell & Bain Ltd., Glasgow

MIX
Paper from responsible sources
FSC® C007785

Contents

List of Tables

List of Figures

Acknowledgements

Thanks are due to the following people who formed the Project Steering Group that provided guidance throughout the entire project: Hoyt Battey (US Department of Energy), Matthew Kennedy (SEAI Ireland), Mette Cramer Buch (Danish Energy Agency), Melanie Nadeau (Natural Resources Canada), Steffen Nielsen (Permanent Delegation of Denmark to the OECD), Michael Paunescu (Project Chair, Natural Resources Canada), Kristian Petrick (RETD Operating Agent, All Green Energies, Spain), Kjell Olav Skjølsvik (Enova Norway), Dorthe Vinther (Energinet Denmark).

Preface

The Implementing Agreement on Renewable Energy Technology Development of the International Energy Agency (IEA–RETD) has appointed Mott MacDonald (MM) to support its role of assisting policy-makers and project developers to better understand the specifics of offshore renewable energy and to give them practical guidelines on how to foster its deployment. The report focuses on the ten IEA-RETD member countries (Canada, Denmark, France, Germany, Ireland, Italy, Japan, The Netherlands, Norway and the UK), as well as eight other countries which have shown activity in the marine renewable industry (Belgium, China, Finland, Portugal, Spain, Sweden, Taiwan and the US).

World electricity production continues to increase and in 2009 reached approximately 17,000 terawatt hours per year (TWh/year). The electricity production process is creating a significant burden on global resources and renewable energy technologies are increasingly part of the mix to meet the challenge of rising energy demand whilst minimising negative environmental impacts.

The world theoretical resource from offshore renewables (wind, wave and tidal) is estimated to be between 260,000 and 330,000 TWh/year, illustrating the potential significance of the available resource. The opportunity to harvest this vast resource has been identified by governments and academia, together with commercial project and technology developers, who aim to capitalize in a rapidly expanding market.

Of all the marine technologies, offshore wind is the front runner, with projects operating since the early 1990s. Even though the offshore wind sector is experiencing high growth, the industry is far from mature and big challenges lie ahead, with projects being planned for deployment further offshore, in deeper locations, with larger machine and technological advancements.

For the wave and tidal sector, a large number of devices are under development, with no particular design having yet emerged as clear front runner. The various technologies are at different stages of development with some prototypes currently being tested at full scale and commercial projects expected in the near future.

Due to the harsh and difficult-to-access environment in which these devices have to be installed and operated, the associated risks (technical and non-technical) are higher than for onshore technologies. Complete removal of such risks is not feasible; however, mitigation measures can reduce these risks to an acceptable level to facilitate project development.

Costs

Regardless of the mitigation measures, the costs of marine projects remain high and uncertain, resulting in financing of a marine project being the biggest barrier for their deployment.

Offshore wind is currently the cheapest of the marine technologies in terms of cost of energy for an installed project (including transmission connections to the shore), with a range of €120–250 per megawatt hours (MWh). Costs of energy estimated for near future pre-commercial wave energy projects are in the region of €140–530/MWh, whereas for tidal energy arrays these costs range from €110–220/MWh installed. Cost of energy for commercial wave and tidal devices should be treated with caution as the uncertainty inherent in these estimates is very high.

Financing

There are a number of financing options available for projects developed by the private sector which are primarily balance sheet and project finance. Each option has its benefits and drawbacks. The characteristics of individual projects and their sponsoring organizations typically dictate which one of these financing options is best suited for a particular project. Balance sheet finance using debt raised corporately is cheaper, involves fewer parties, and control of the project remains firmly with the owner; it is, however, capital intensive and the risk of failure lies entirely with the owner. On the other hand, project finance allows greater leverage from the available funds for sponsors' equity investment; however, it is typically more expensive and complex and an element of control over the project is afforded to the lenders. One solution is to finance construction projects on balance sheet and move to project finance on completion, recycling development capital into new projects.

To date, wave and tidal stream projects have not been project financed. With the most advanced technologies typically at a pre-commercial/prototype stage, they are seen as containing large amounts of technology and performance risks. Funding for technology deployment to date has tended to tap venture capital or public-sector development support sources. Project developments are mainly pursued by utilities. A project finance model may emerge in the future once the technologies have been de-risked.

Tariff support

The lack of long-term or stable policy commitments from governments is another significant barrier as it affects developer and market confidence. Furthermore, in many countries, the level of financial support provided (feed-in tariff or tradable certificates) often appears either insufficient or, at best, marginal in order to

Figure 0.1 Wave energy device
Source: Aquamarine Power www.aquamarinepower.com

provide sufficiently attractive returns to investors compared to lower risk investment options in other sectors.

Confidence in long-term market opportunities is required from the private sector in order to trigger the investment decisions necessary to the development of a supply chain for the offshore renewable sector. At the national level, governments can heavily influence and coordinate the development of required infrastructures (such as harbours and grid). The importance of public support for marine technologies is illustrated by the success that offshore wind has had in some countries where such a barrier was removed. Countries that have shown strong political leadership and tailored financial incentives, such as Belgium, Denmark, Germany and the UK, are leading the way in terms of deployment.

Planning and permitting

Complex permitting processes are another major barrier to offshore renewable energy project development in most countries. Prescriptive planning conditions or requirements limit projects and technology design options and can significantly increase timescale and development costs. A number of regulatory barriers are also delaying or preventing the changes required in the onshore and offshore grid infrastructure in order to accommodate offshore (and onshore) renewable expansion plans. Deployment timescales can be greatly increased as a result of these barriers.

While the permitting processes are diverse and country specific, lessons can be learned from the countries that have had more success with offshore wind. Streamlined application procedures, one-stop shops and pre-permitted areas are some of the potential mitigation measures to planning and permitting barriers. The allocation of seabed rights to competent and construction-focused developers is also important in order to avoid sites being leased to developers more interested in speculative applications or without the necessary resource to progress the development of projects.

Technologies

Offshore wind turbine technologies are based on the three-bladed upwind horizontal axis design, although new concepts are being developed. Offshore wind foundations can be split into six broad categories (monopile, multipile, gravity base, jacket, suction cup and floating foundations). Their suitability is mainly linked to water depth and seabed conditions.

Wave and tidal energy technologies are at a much earlier stage of development compared to offshore wind. A wide range of technologies and designs are currently being developed. The main tidal energy designs include horizontal axis turbines, vertical axis turbines and oscillating hydrofoil. Wave energy designs are more diverse and main categories include attenuators, point absorbers, oscillating water columns (OWCs), oscillating wave surge converters and overtopping designs.

Floating marine energy devices require mooring systems, for which there are various designs available or under development.

The main technical challenges and barriers shared by all marine renewable energy technologies include technology and design optimization, reliability, installation and decommissioning, operation and maintenance, grid connection, and integration. Considerable investments will be required in onshore and offshore grid infrastructure in order to accommodate the large expected expansion in variable generation capacity from offshore renewable energy projects. In some parts of the world, the optimal topology of this expansion needs to be considered at a supra-national rather than national level. Technical barriers are surmountable, but usually impact the cost of offshore renewable energy project and technologies.

Research, development, demonstration and green employment

Research, development and demonstration (RD&D) activities performed directly by the private sector or financially supported or promoted by public funding are instrumental in the removal or mitigation of technical barriers, and through

Figure 0.2 Three-bladed horizontal axis turbine: Offshore wind turbine at EnBW Baltic 1 offshore wind farm
Source: EnBW www.enbw.com

Figure 0.3 Seagen horizontal axis tidal turbine
Source: Marine Current Turbines

creating domestic intellectual capital can also support green employment and the development of future industries. The importance of the support that can be provided by publicly funded RD&D activities is particularly relevant for the more immature technologies given the lower investment capacity of the private sector and longer timescales involved. Direct involvement and possibly co-investment from private companies into RD&D activities should be maximized.

Other barriers

Other barriers include health and safety, environmental and other sea users' considerations, supply chain constraints and skills shortages. While all of these issues are important and have to be dealt with, mitigation measures can reduce their impact.

The reduction or mitigation of health and safety barriers can be achieved by a strong industry culture, supported by staff training, compliance with legislative requirements, best practices and standards, as well as through technical innovations.

The main mitigation measures in order to ensure the minimization of environmental barriers and acceptance from other sea users are early engagement with all stakeholders, appropriate marine spatial planning and adoption of the recommendations from the environmental impact assessment (EIA).

Figure 0.4 Substation installation by floating crane at EnBW Baltic 1 offshore wind farm
Source: EnBW www.enbw.com

The infrastructure, products and services supply chains need to be vastly developed in order to increase competition and avoid shortages. This can be delivered by private-sector investment, but only if governments establish sufficient confidence in the long-term market opportunities.

The removal of skills barriers requires the active promotion of the various employment and careers opportunities provided by the offshore renewable energy industry, as well as the development of training courses and programmes tailored to the needs of the industry.

Project development

Developers are recommended to follow best practice at all stages of projects' lifecycle, from pre-feasibility, development, design, construction, operation and decommissioning.

Examples of such best practices include definition of clear project objectives and development strategy, early consultation with other stakeholders, and creation of a risk register to be maintained as a live document throughout the project life as a tool to record and then address the key project risks.

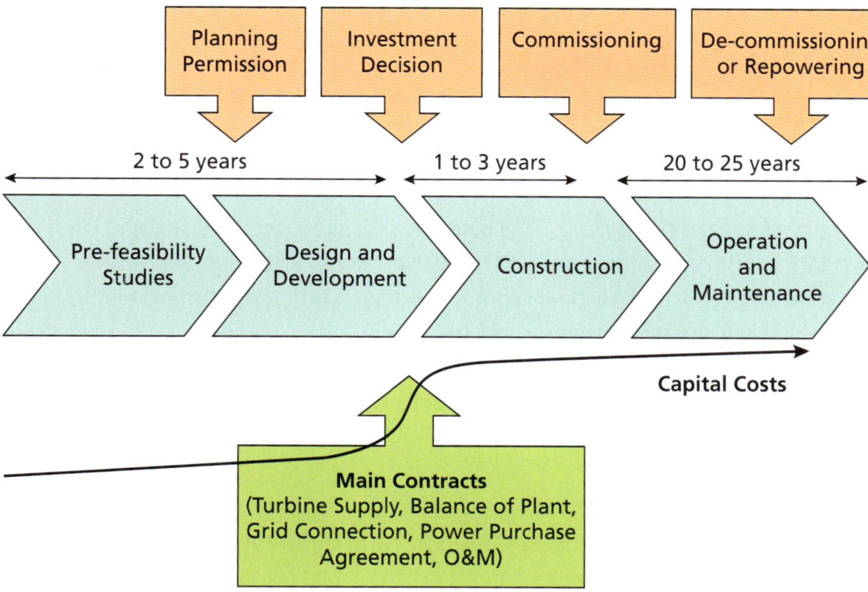

Figure 0.5 Lifecycle of wind turbine project

Model policy framework

A conceptual model policy framework that promotes offshore development needs to include strategic support mechanisms, stable regulatory regimes, efficient permitting and grid connections, and access to finance. Such a framework also needs to include measures supporting innovation and competition. The support mechanisms will have two distinct roles:

- create an orderly environment in which developers can work; and
- de-risk this new industry.

The latter mechanisms will need to be formulated so they can be diminished or withdrawn as experience is gained to reduce costs to taxpayers or electricity consumers, and give developers rewards proportionate to their risks.

The proposed model policy framework should include the following aspects:

- *Market creation*. Provide flexible support mechanisms appropriately sized for the size of risk undertaken and the overall desired capacity to be derived from offshore renewables. This may take the form of:
  - 'phased' tariffs, whereby initial prototype/pilot projects are implemented in relatively calm waters to enable the industry to go through a lessons

  learned process, with later projects receiving less support once learning has occurred; and/or
  - tariffs that are flexible depending upon the offshore resource, water depth and distance from shore of particular sites; or
  - a tendered capacity model where tariffs are bid to develop projects at particular sites.
- *Straightforward permitting*. This involves one-stop agencies instead of engagement with a large number of government agencies. Permitting requirements should be clear from the outset. Define offshore development zones where offshore projects can be developed.
- *Grid connection facilitated*. Clear arrangements to provide the necessary grid connection (onshore and potentially offshore) in a timely manner, with adequate commercial recourse should grid connections not be available in time.
- *Early development supported and de-risked*. Offshore resource measurement campaigns, seabed surveys and other measures in areas of interest, such that developers can assess the basic feasibility of project investments. As the industry matures, the level of such support can be reduced over time. Licences should include expiry dates and require clear achievement of milestones from the developers in an effort to minimize sites being reserved for projects that will not materialize.
- *Access to capital*. Fund pilot projects with the industry (demonstration projects, measuring campaigns, etc.). Create support mechanisms aimed at providing early investment (grants for device development, tax relief, financing for projects, underwriting of a project, etc.). Other measures could include the creation of a government financing body for projects to support commercial bank financing and provide a further signal to the lending community that the government is strongly supporting this industry. These support mechanisms will diminish as the industry begins to mature.
- *Supply chain creation (for countries wishing to provide a supply chain)*. Provide funding for device, foundations, mooring and other peripheral services development and support research in the relevant field. Provide suitable manufacturing bases and suitable harbours with further support (tax breaks, etc.). Create centres of excellence and strongly market to the rest of the world. Support conferences, seminars and other forms of networking and knowledge transfer.
- *Skills development*. Identification of shortage skills and a programme tailored to their proactive development, as far as possible, over the required timescales.
- *Clear environment, health and safety legislation*. Outline clear environmental requirements in line with the Equator Principles. Adopt strong internationally accepted health and safety (H&S) guidelines.

The policy framework should receive strong visible support from government and government organizations to emphasize the commitment to the industry. Nevertheless, once the projects are in the construction phase, government intervention should be kept to the minimum.

Figure 0.6 Bremerhaven Port current facilities and planned extension for offshore wind
Source: offshore-windport.de

1 Introduction

Mott MacDonald (MM) has been appointed by the International Energy Agency's Implementing Agreement on Renewable Energy Technology Development (IEA–RETD) to support its role of assisting policy-makers and project developers to better understand the specifics of offshore renewable energy and to give them practical guidelines on how to foster its deployment. This book, the deliverable of this appointment, is broken down into four parts:

1. overview of the industry context (Chapters 2 to 4);
2. economics and financing of offshore energy projects (Chapters 5 to 7);
3. technical and non-technical barriers and their mitigation measures (Chapters 8 and 9), as well as guidelines for project development (Chapter 10);
4. summary of findings and conclusions, leading to recommendations and next steps (Chapter 11).

Throughout the book, case studies and examples of existing, as well as planned, offshore renewable energy projects are used to expand upon specific key issues described.

In Appendix E more detailed profiles of the countries are included, discussing the resource assessment, support mechanisms and deployment to date in each of these countries.

1.1 Industry context

Chapters 2 to 4 set out and summarize the industry context. They outline assessment of offshore resources, technology briefs and deployment in selected countries and status of policy development and technology deployment. The aim is to set the scene of the industry and provide a platform for the analysis undertaken in the rest of the book.

The industry context is split into three main chapters:

- 'Offshore resource' (Chapter 2);
- 'Offshore renewable energy technologies' (Chapter 3); and
- 'Deployment targets, policies and progress' (Chapter 4).

The layout is designed to provide an appreciation of the resource potential and justification for allocating resources to exploit it. As such, Chapter 2 provides an overview of global resources, highlighting the potential benefit of exploiting such resources. Chapter 3 then follows by identifying technologies that can be used to utilize such resources, aiming to prove that it is technically feasible to do so. Policies and targets are presented in Chapter 4, setting the regulatory framework for facilitating such projects, followed by an overview of the deployment of the technologies and their success to date.

The study focuses on the experience of 18 countries: 10 RETD countries and 8 other countries of interest. These are listed in Table 1.1. More detailed country profiles discussing the resource assessment, support mechanisms and deployment to date for each of these countries are presented in Appendix E.

Table 1.1: Countries covered by the study

RETD Countries	Other Countries of Interest
Canada	Belgium
Denmark	Finland
France	Spain
Germany	Sweden
Ireland	US
Italy	China
Japan	Taiwan
The Netherlands	Portugal
Norway	
UK	

1.2 Economics and financing of offshore energy projects

Chapters 5 to 7 aim to improve policy-maker understanding of the cost structure of offshore renewable energy projects. They cover the following topics:

- 'Economics of offshore energy projects' (Chapter 5);
- 'Project risks and related project costs' (Chapter 6); and
- 'Financing of offshore renewable energy projects' (Chapter 7).

Chapter 5 introduces the concept of technology maturity and technology readiness levels (TRLs) and goes on to compare the capital expenditure (CAPEX), operational expenditure (OPEX) and cost of energy of the different technologies, including the cost structures and drivers and variations by country. Chapter 6 considers how project risk assessment affects the economics of projects, while Chapter 7 looks at financing options, including balance sheet finance and project finance.

1.3 Technical and non-technical barriers and their mitigation measures

Chapters 8 and 9 aim to improve policy-makers' and project developers' understanding of the various technical and non-technical barriers and challenges faced by offshore renewable energy projects and how these affect the overall cost structure and development timescales. For each type of barrier, the book goes on to describe how policies and other measures can be developed and implemented to remove, reduce and overcome the challenges faced. Based on the findings of the previous sections, Chapter 10 provides a set of guidelines for the development of offshore renewable energy projects.

Chapter 8 analyses non-technical barriers and their mitigation measures. This includes those common to all offshore renewable energy technologies and those specific to offshore wind and to wave and tidal separately. Grid connection barriers are also considered in depth. Chapter 9 focuses on the main types of non-technical barriers – namely, environmental; health and safety; regulatory and permitting; competing use; skills availability; supply chain and infrastructure, and access to capital and financial support mechanisms. Barriers can differ from country to country and from one technology to another. The book describes general and, where applicable, country- and technology-specific barriers.

1.4 Findings, conclusions, recommendations and next steps

Chapter 11 summarizes the findings and conclusions, leading to the recommendations and next steps. A clear policy framework can play a vital role in securing investment in offshore development and, over time, help to support innovation, a competitive environment and reducing costs. The recommendations therefore include a model policy framework.

2
Offshore Resource

2.1 Introduction

This chapter provides a summary of the mechanics of wind, wave and tidal energy and an outlook of the world offshore energy potential. The chapter includes a table of the offshore resources for each of the countries covered by the study.

2.2 Resource assessment

Resource assessments are based on measurement data and assumptions. Commonly, four categories are referred to and are differentiated as follows:

1. Theoretical potential determines the entire energy resource physically available.
2. Technical potential determines the maximum amount of energy that can be captured using the technical means available.
3. Practical potential additionally takes into account external constraints (e.g. competing use of land and sea surface or environmental sensitivity).
4. Economic potential considers, in addition to technical limits and external constraints, economic drawbacks.

2.3 World distribution of offshore resources

2.3.1 Offshore wind energy

Wind is a directly derived form of solar energy. Solar radiation on the Earth's surface results in the warming of the atmosphere, water and land masses. Temperature differences occur due to the varied surface structures of the Earth, day and night cycle, and abundance of solar irradiance near the Equator compared to the poles. These temperature differences result in pressure differences and set the air masses into motion. Additionally, the Earth's rotation contributes to the turbulence of the air mass.

Wind energy is a function of four factors:

1 density of the air passing through the rotor, ρ;
2 wind speed, v;
3 swept area, A; and
4 power coefficient, cp (i.e. theoretical limit of power conversion – approximately 0.59).

As the following formula illustrates:

$$P_{Rotor} = \frac{1}{2} \cdot \rho_{Air} \cdot A \cdot v_{wind}^{3} \cdot c_p$$

It is important to note that in this equation wind speed is cubed, resulting, for example, in eight times the energy output with double the wind speed.

The wind force in lower atmospheres depends upon the roughness of the surrounding area. Generally, friction slows wind down. Over open waters, friction is particularly low, favouring the development of strong winds. Figure 2.1 shows the worldwide mean wind speed, highlighting that the wind speed is highest over the open oceans. Archer et al estimated a global wind resource of around 72TW.[1]

Wind energy has been utilized for a number of years for electricity production onshore. However, the realization that the wind resource is favourable offshore, as well as technical advancements, led to the first offshore projects being developed during the 1990s. Since then, significant growth has been exhibited and current projections for Europe alone estimate a substantial technical potential for offshore wind energy development until 2020.[2]

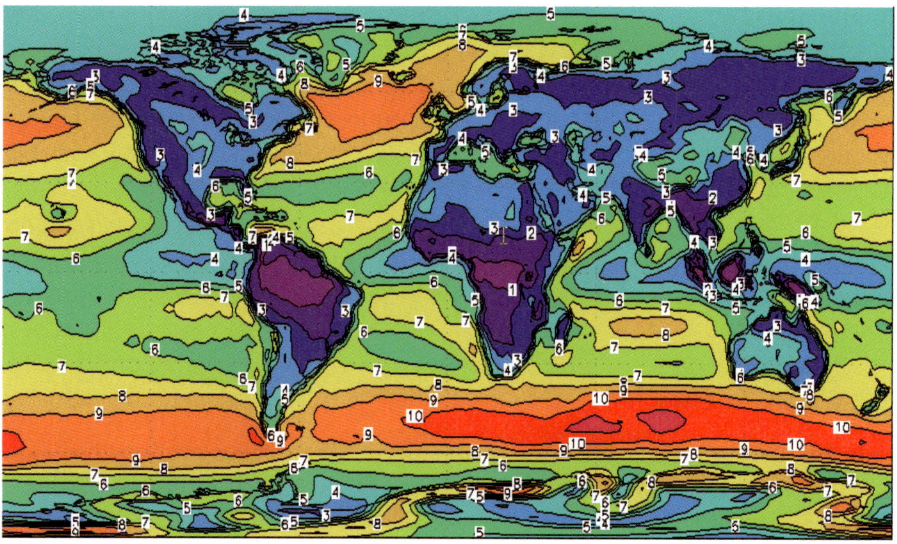

Figure 2.1 Mean wind speed (in m/s) for 1976–1995 according to the NCEP/NCAR Reanalysis Data Set

Source: Risø National Laboratory for Sustainable Energy

2.3.2 Wave energy

Wind blowing over open water surfaces, such as oceans, creates waves by transferring part of the wind's energy. Surface ocean waves represent a concentrated form of wind energy, with the wave height determined by the following main factors:

- wind speed;
- duration for which the water is exposed to the wind;
- open water fetch (distance over which waves are exposed to wind);
- water depth; and
- topography of sea bed.

The energy contained in waves is commonly referred to as the energy flux per metre of crest length (wave front). For waves in deep water, the energy flux is defined by the significant wave height Hs and the wave period T as per the following formula:

$$P_{Rotor} \cong \frac{1}{2} \cdot H_S^2 \cdot T$$

The theoretical resource is estimated between 8000 and 80,000 TWh/year.[3,4] Values between 145[5] and 2000TWh/year[6] are assumed for the economically or practically extractable resource. Carbon Trust estimates for practical worldwide resource a range between 2000 and 4000TWh/year.[7] The range of potential wave resource is very large. The main reason is the lack of (or no) data for several countries, as can be seen in Appendix E. Some studies have also not been very concise. The *World Wave Atlas* produced by Fugro Oceanor presented in Figure 2.2 shows the distribution of wave energy.

For Europe, extensive research has been undertaken and the wave energy resource is estimated to be in the region of 320 gigawatts (GW).[8] The available wave energy resource for the north-eastern Atlantic, including the North Sea, has been estimated to be about 290GW, with wave power levels ranging between 25 and 75 kilowatts per metre (kW/m). An annual deep-water resource of the European coast of the Mediterranean Sea has been assumed to be in the order of 30GW.

2.3.3 Tidal energy

Tidal energy is generated from a combination of the relative motions between Earth, Moon and Sun and the Earth's rotation. The changes in the gravitational forces induce cyclic variations of the sea level, which result in water currents. The rise and fall of water levels is predominantly in the deep ocean, from where

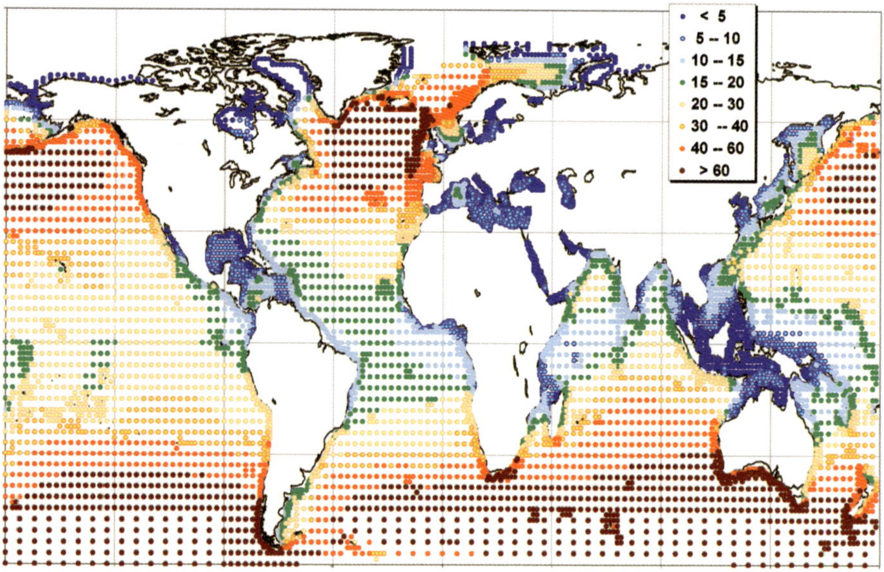

Figure 2.2 Mean annual wave power (*World Wave Atlas*)
Source: Fugro OCEANOR and ECMWF (European Centre for Medium-Range Weather Forecasts)

they propagate as a long period wave (12.4 hour wave length). Tides are therefore experienced as height difference between low and high tide (referred to as tidal range) and the tidal stream. Both features represent a form of energy, and various technologies to harvest these are under development.

The energy potential of tidal currents is a function of the fluid density ρ, the swept area A and the local fluid velocity U as per the following formula:

$$P_{Turbine} = \frac{1}{2} \cdot \rho \cdot A \cdot U^3$$

The principle of this calculation is similar to wind energy and highlights the importance of the higher density of water compared to air.

A global assessment of the tidal resource has not been undertaken yet; but the following estimates have been made:

- worldwide tidal stream resource of 180TWh/year;[9]
- theoretical world resource of 365TWh/year;[10]
- tidal currents above 800TWh/year;[11] and
- for the European tidal stream resource, a figure of 13.7GW has been suggested.[12]

Based on satellite data, global models of ocean tides have been developed. Figure 2.3 provides an impression of areas with high tidal variability. Specifically, the map shows the global distribution of frictional energy dissipation due to tidal currents.

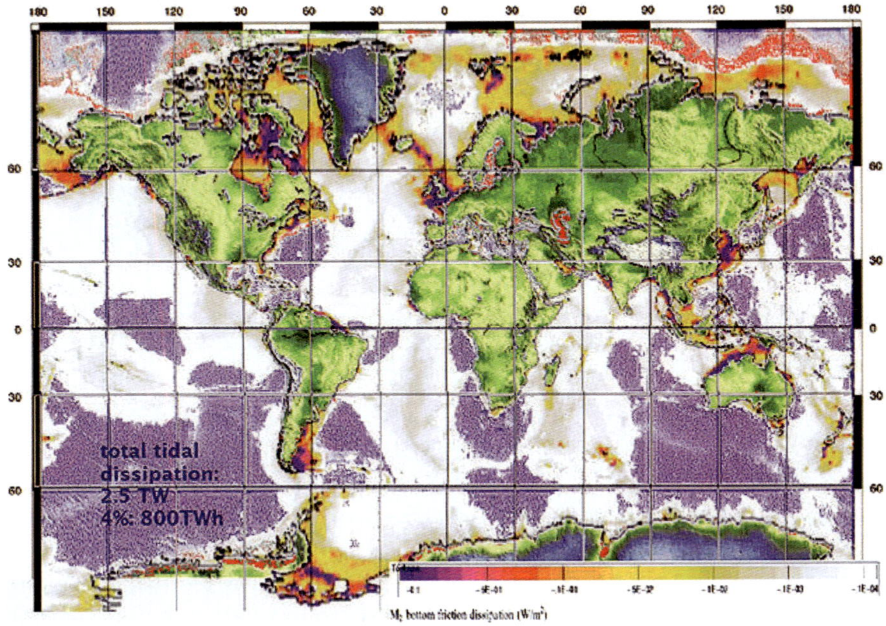

Figure 2.3 Tidal resource model
Source: Lyard et al (2006)

A number of tidal models exist and have been utilized to create maps of tidal currents. The resolution of the available models tends to be insufficient to obtain reliable predictions of tidal flows near the coast, particularly at high-energy sites. Generally, it is recognized that favourable sites should reach a current velocity at peak of spring tides of 2.5m/s.[13]

2.4 Summary of offshore energy resources by country

Table 2.1 provides a summary of the selected countries as indicated in section 2.1. Details of the offshore resource for the illustrated countries can be found in Appendix E ('Country profiles'). The world electricity consumption in 2009 accumulated to 17,130TWh.[14] It is evident from Table 2.1 that, theoretically, offshore energy is capable of supplying such demand.

The data provided refer to the theoretical potential of the resources. The lack of reliable information and data limits the reliability of resource assessments. Some assessments of the technical, practical or economic potential of the

Table 2.1 Offshore energy resources by country

Country	Wind Resource (TWh/year)	Wave Resource (TWh/year)	Tidal Resource (TWh/year)
World	~252,300[1] *	8000–80,000[2]	180[3], 365[4], 800[5]
Canada	n/k	~640[6] **	~185[7] ***
Denmark	~26[8]	30[9]	n/k
France	10[10]	98[11]**	n/k
Germany	20–200[12]	n/k	n/k
Ireland	11[13]	187[14]–525[15]	n/k
Italy	10[16]	n/k	n/k
Japan	10–413[17]	n/k	n/k
The Netherlands	33[18]	n/k	n/k
Norway	14,000[19]	~400[20]	n/k
UK	230–334[21]	~420[22]**	18[23]
Belgium	1.2[24]–4[25] *	n/k	n/k
Finland	20[26]	n/k	n/k
Spain	7[27]	n/k	n/k
Sweden	22.5[28]	n/k	n/k
US	~4100[29] *	2100[30]	13[31]

Notes: * 40% capacity factor assumption.
** 50% capacity factor assumption.
*** Includes tidal and ocean current resource.
n/k = not known from published sources identified for this study.

Sources: *1 Archer et al (2005)*
2 IEA (2006)
3 Black and Veatch (2005)
4 Canadian Hydraulics Centre (2006)
5 IEA (2006)
6 Canadian Hydraulics Centre (2006)
7 Canadian Hydraulics Centre (2006)
8 Duwind (2001)
9 Marine Institute (2002)
10 Duwind (2001)
11 Clément et al (2002)
12 BWE (2010)
13 Duwind (2001)
14 Clément et al (2002)
15 Marine Institute (2005)
16 Duwind (2001)
17 WEC (2007)
18 Duwind (2001)
19 Enova (2007)
20 Marine Institute (2002)
21 Duwind (2001)
22 Thorpe (1992)
23 Black and Veatch (2005)
24 Duwind (2001)
25 Van Hulle et al (2004)
26 Duwind (2001)
27 Duwind (2001)
28 Duwind (2001)
29 Duwind (2001)
30 Duwind (2001)
31 Duwind (2001)

resources have been undertaken for specific countries or locations and are mentioned in the country sections in Appendix E. Overall, the need for further research of resources in some of the listed countries in order to increase the accuracy of the potential should be noted. For any wind, wave or tidal project, site-specific measurements will be required from prospective developers in order to establish the exact resource at the site of interest. Project development issues are dealt with in latter sections of this book. In deploying very large amounts of offshore energy as a proportion of national or system demand, there would also most likely be practical issues in matching it with the pattern of energy demand, and the availability or otherwise of a storage capability.

Offshore wind energy seems to be a valid option for all countries in the focus of this book, with Norway, the US and China featuring the highest resource figures. Wave and tidal resources give a different picture since some of the countries lie in sheltered areas of the oceans. Countries with a particularly high wave resource include Canada, Ireland, Norway, the UK and the US. Finally, high tidal resource can be found in Canada, Ireland, the UK and the US.

Notes

1 Archer et al (2005).
2 EEA (2009).
3 AEA/SEI (2006).
4 IEA (2006).
5 Wavenet (2003).
6 Thorpe (1999).
7 Carbon Trust (2006).
8 Clément et al (2002).
9 Black & Veatch/The Carbon Trust (2005).
10 Canadian Hydraulics Centre (2006).
11 IEA (2006).
12 AEA/SEI (2006).
13 Farber Maunsell/Scottish Executive (2007).
14 CIA (2010).

3

Offshore Renewable Energy Technologies

In the following sections, brief presentations of the three main types of offshore renewable energy technologies (offshore wind, wave and tidal) are provided, as well as associated structures such as foundations, mooring and electrical infrastructure elements (substation, cables).

3.1 Offshore wind devices

The offshore wind industry is by far the most developed of all offshore energy industries, with commercial units deployed in a number of countries. As such, there is reliable information based on realized projects and operational data in contrast to wave and tidal, where most of the information is based on projections of researched designs.

Nevertheless, while offshore wind is currently enjoying high growth, the industry is still not considered mature. With increasing distance from shore and water depths exceeding 40m, there are some additional technical challenges which need to be overcome.

3.1.1 Working principle

The fundamental working principle of a wind turbine or, more specifically, a wind energy converter is that it utilizes the means of aerodynamic profiles to convert the kinetic energy stored in wind into electrical energy. In this principle, the offshore variants are no different to onshore models. While a number of designs have been used, to date, only one system has been commercially successfully implemented onshore and offshore: the three-bladed upwind horizontal axis wind turbine. While there are other designs currently being developed, such as vertical axis and two-bladed upwind horizontal axis wind turbines, this book focuses on the horizontal version as this design is expected to dominate the market for the next ten years.

The turbine captures the wind energy with three blades which, in principle, work like a propeller in reverse. The wind runs past the blades in a near horizontal direction and forces the aerodynamic wings of the turbine to rotate. This rotational energy is typically transferred via a shaft to a gearbox to increase the relatively slow rotation of the blades (approximately up to 20 revolutions per minute – rpm) to a more generator-compatible rpm range (typically about 3000/3600 rpm). The gearbox is connected to an electrical generator that converts the kinetic energy into electrical energy, which is then transformed to a higher voltage to be suitable for grid injection.

While the basic operational principle of the offshore turbine is no different than onshore, alterations are required to make it suitable for deployment offshore. The corrosive environment offshore and the high levels of moisture in the air would lead to electrical and mechanical problems if an onshore design were used without modifications. Therefore, modern offshore turbines use air-conditioning systems to protect the sensitive electronics inside the unit and protective paint to protect the steel structures. These systems work well under the maintenance regime used by the operators.

3.1.2 Offshore wind turbine manufacturers

Currently, there are a number of offshore wind turbines available on the market, with Siemens being the current market leader with its 2 and 3MW class units. Vestas is, however, recovering from its initial problems with its offshore units and has introduced a new 3MW design which is highly attractive to the offshore market.

An overview of the models that have been deployed offshore and are currently available to the market can be found in Table 3.1. While the table shows that a large number of 2MW class units have been deployed offshore, current and future projects will utilize bigger units over 3MW in order to maximize profits.

Some manufacturers are currently producing large-capacity wind turbines while keeping the rotor diameter near constant to minimize wake effect. There are also manufacturers developing turbines of the 10MW class that have a similar footprint to the currently available 5MW units. While ambitious projects such as the UpWind project are aiming at rotor diameters beyond 200m and output in the 20MW[1] range, these units are in the early stages of development. Direct-drive turbines with no gearbox are expected to be more prominent in the future. The Vänern lake offshore wind park, where ten WWD-3 wind turbines were installed and which was inaugurated in May 2010, represents the world's first inland offshore project.

According to a European Wind Energy Association (EWEA) report,[2] the European offshore market has 3000MW of installed capacity. The expected amount of offshore wind energy installed in 2010 was exceeded and amounted to 1444 MW; therefore, the cumulative wind capacity installed offshore currently amounts to 3554MW (BTM Consult Aps, 2011). With a planned expansion to

Table 3.1 Installed offshore turbine models

Manufacturer	Model	Output	Units installed offshore
Siemens	SWT 2.3–93m	2.3MW	136[1]
Siemens	SWT 3.6–107m	3.6MW	103[2]
Vestas	V80	2MW	120[3]
Vestas	V90	3MW	126[2]
Multibrid/Areva	M5000	5MW	6[2]
REPower	5M	5MW/6MW	8/0[2]
Bard	VM	5MW	12[2]
Sinovel	SL3000	3MW	34[4]
WinWind	WWD-3	3MW	10[4]

Sources: 1 Siemens (2009); 2 WAB (2009); 3 Vestas (2010); 4 BTM Consult Aps (2010)

40,000MW by 2020, it is clear that this target can only be achieved with bigger capacity units.

Many of the turbines listed in Table 3.1 are evolutions from onshore turbines with only the REpower 5M, Multibrid/Areva M5000 and Bard VM being designed specifically for offshore usage. Although these machines have been designed with offshore application as a prime design driver, approaches have been different – for example, M5000 is based on the concept of a simple design with built-in redundancies, while others have focused on extending the service intervals. It is anticipated that all future models of the 5MW+ class will be specifically designed for offshore environments.

3.1.3 Key components

A wind turbine is comprised of several key components, as shown in Figure 3.1 – namely, blades, hub, main shaft, main bearing, nacelle, gearbox, generator, yaw section, as well as the tower and ancillary items, which are not shown on the illustration. A more detailed explanation of these components is given in the following sections.

Blades

Current state-of-the-art 5MW class wind turbine features blades with lengths exceeding 60m, about the same as the wingspan of a Boeing 747, weight in excess of 17 tonnes and a root diameter of over 3m. Each blade is attached to the hub with over 100 bolts of significant size, typically 36mm in diameter (twice

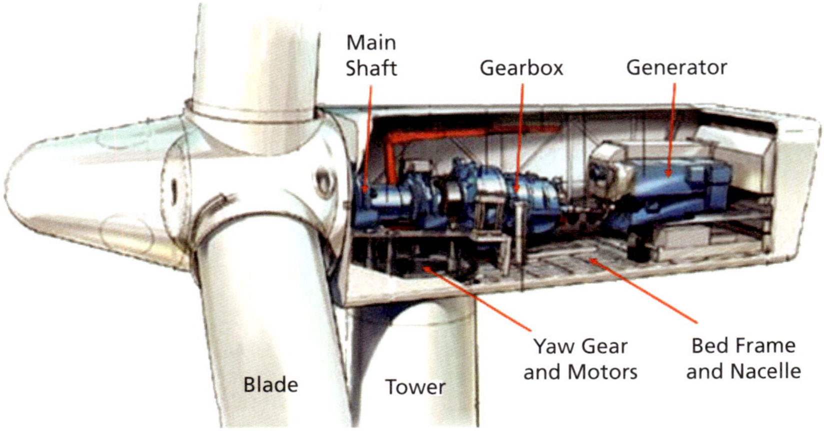

Figure 3.1 Schematic of typical wind turbine layout
Source: Siemens AG

that of a wheel lug on a car). Each blade is made of a proprietary mixture of man-made materials and is shaped specifically to each manufacturer's aerodynamic specifications. In addition, each blade is equipped with lightning-strike conductors (lightning protection).

Each blade is designed and certified to withstand at least 20 years of operation and is subject to rigorous tests to confirm that it can handle the dynamic loads imposed by the rotation of the blades, and centrifugal and wind forces. The blades of a wind turbine are by nature an essential component to the operation of the entire system; a failure would not only interrupt the operation, it would typically also cause a large amount of collateral damage to the remaining components, including the tower, remaining blades and hub.

Hub

The hub is the assembly that joins the blades together to form the rotor, which transfers the rotational energy to the gearbox via the shaft. The hub is most commonly a large piece of cast metal and houses a variety of electronics to regulate the pitch (the angle of the blade to the wind) of the turbine in order to most efficiently extract the energy from the wind. As it is a cast piece, careful consideration needs to be given to the casting process when this product is designed in order not to restrict the supply chain and hinder mass production.

Typically, a heavy protective paint coating is applied on all surfaces to protect the hub from the hostile environment and prolong the maintenance interval to at least 15 years. In addition, several designs incorporate an air-conditioning system which creates an overpressure of dry and salt-free air on the inside of the hub to prevent ingress of the corrosive environment.

Depending upon the design, the hub is bolted directly either onto the gearbox or to the main shaft.

Main shaft

The main shaft, if part of the design, transfers the high-torque low-speed revolutions into the gearbox and can vary in design from a hollow to solid shaft. It supports the entire weight of the rotor and transfers any loads into the main bearing. Typically it is equipped with several slip rings to transfer data signals into the hub and to transfer potential lightning strikes away from the blade and hub into the main frame of the nacelle.

Main bearing

The main bearing allows loads, such as gravitational and rotational forces, to be transferred into the main frame and by doing so prevents these forces from being transferred into the gearbox where they would potentially cause damage. Axis loads are, however, not absorbed here; they can normally be compensated for by the gearbox itself (depending upon the design).

Nacelle and main frame

The nacelle houses the main components of a wind turbine such as the gearbox, generator, cooling system, electronic controls and transformer (depending upon the design). Most nacelles today are made from a steel frame (the main frame) covered with glass fibre. This design yields a weight reduction while maintaining the required structural integrity. All loads imposed by the rotor are contained by the main frame and transferred into the tower from which they are transferred into the foundation of the turbine. It is essential that the main frame is designed in such a way that the transferred energy does not cause it to deform as this would cause misalignment in the system, which reduces the lifetime of the wind turbine. Conventional nacelle layouts consist of the following components (listed in reverse direction from the rotor): main bearing, main shaft, gearbox, generator and cooling system. Around the periphery of the nacelle, manufacturers typically position the electronic control systems.

Gearbox

The gearbox converts low-speed high-torque rotational energy into high-speed low-torque rotational energy, which can be passed on to the generator. There are many designs; however, most turbines currently in operation use a 'conventional' design consisting of multiple stages of planetary and helical gears. It should, however, be mentioned that some newer turbine designs use hybrid designs with no gearbox or with a gearbox of reduced complexity. These designs have not yet featured in offshore wind *en masse*; however, the gearless design is viewed favourably by the industry.[3]

Generator

Generators in wind turbines have one function: converting rotational energy into electrical energy with the least amount of losses. Again, there are several designs in the marketplace at the moment. Most state-of-the-art designs use a double-fed induction generator; however, permanent magnet generators are appearing with increasing frequency in new designs.

Yaw section

The yaw section of the wind turbine is comparable to a weather cock and ensures that the turbine is always headed into the wind. This is achieved by use of electric motors and a gear ring on the top of the tower.

Tower

Offshore towers are treated with a corrosion protection coating on the outside to maintain their integrity over the 20-year or more life time. Commonly, it also houses part of the air-conditioning system, which provides dry and virtually salt-free air to the interior of the turbine.

Air conditioning

It has become customary over the past years to include an air-conditioning system in offshore wind turbines. By providing dry and salt-free air to the interior (and sometimes the hub) of the turbine, many manufacturers have been able to reduce the requirements on their electronic components as these are now operating in a quasi-onshore environment. This is beneficial not only for the electronic components, but also for any metal components which would otherwise be exposed to the environment. Some manufacturers, however, do not think that air conditioning is required.

3.1.4 Deployment location

Technical factors influencing the deployment of offshore wind turbines are the distance to shore, the water depth and soil conditions. Section 3.4 provides an overview of the available foundation designs accounting for these three parameters.

3.1.5 Costs

Over the past years, the industry has seen a change in the way that offshore wind farms are built. At present, no contractor is willing to take the full technical risk of the project and the developer has been pushed to take on more risk. In addition, the majority of projects currently being developed are being built by a multitude of contractors all working independently of each other. This places a significant amount of coordination work onto developers and leaves them with much of the technical and commercial risk.

As a result, 'early' offshore wind projects were developed at specific costs between €1100/kW and €2000/kW[4] (2001 to 2007). Later projects (2008 to late 2009) have exhibited costs between €2700/kW up to €5100/kW.[5] Turbine prices are currently on the rise as the demand for proven technology turbines to meet the European Union (EU) climate change requirements has increased dramatically and developers are struggling to secure a production share.

3.2 Wave energy devices

This section provides a brief overview of the wave energy technologies that could harness the offshore energy resource. The following technologies are considered:

- point absorber;
- attenuator;
- oscillating water column (OWC);
- pitching/surging/heaving/sway/sloped (PSHSS); and
- overtopping.

A number of attempts have been made to classify marine (wave and tidal) and, specifically, wave energy devices. The methods of classification include references to the combination of motions used by the device and the particulars of the power conversion mechanism. For this assessment, the available devices have been categorized based on their motion as per the guidelines of the Wind and Hydropower Technology Program of the US Department of Energy.

3.2.1 Point absorber

Working principle

Point absorbers are devices that are relatively small compared to the longer wave length in which they operate. Numerous devices at different stages of development fall within this category. Point absorbers generally consist of a buoy which may be surface piercing or sub-surface and a power take-off system which includes, at minimum, a mechanical linkage or hydraulic system and a generator. The buoy moves in an upwards and downwards motion as waves pass the device. Movement of the buoy is coupled to the generator, which converts the kinetic energy of the buoy to electrical energy.

Figure 3.2 Archimedes wave swing
Source: AWS Ocean Energy

Available and researched designs

Table 3.2 summarizes some available and researched designs.

Table 3.2 Point absorber designs

Company	Technology	Device Type	Country Base	Stage of Development
AWS Ocean Energy	Archimedes Wave Swing	Sub-surface point absorber	UK	Pilot tested test deployment of a full-scale CETO III unit was due for completion in 2009, with commercial rollout anticipated shortly thereafter
Columbia Power Technologies	Direct Drive Permanent Magnet Linear Generator	Surface – point absorber	US	10kW units

Company	Technology	Device Type	Country Base	Stage of Development
Fred Olsen & Co./Ghent University	SEEWEC	Surface – point absorber	Norway/ EU	1/3 model of platform tested 2005 40kW point absorber pilot project
Ocean Navitas	Aegir Dynamo	Surface – point absorber	UK	Prototype 100kW unit not tested
Ocean Power Technologies	Power Buoy	Surface – point absorber	UK/US	Prototype tested
Renewable Energy Holdings	CETO	Surface – point absorber	Australia/ UK	Prototype tested
Seabased AB	Linear generator (Islandsberg project)	Surface – point absorber	Sweden	Deployed tested and connected to grid
Trident Energy Ltd, Direct Thrust Designs Ltd	The Linear Generator	Surface – point absorber	UK	Fully functional demonstration WEC due 2010
Wave Energy Technologies Inc.	WET EnGen™	Surface – point absorber	Canada	40kW pilot
Wave Star Energy ApS	Wave Star	Surface – point absorber	Denmark	500kW prototype currently being constructed and tested (2011 completion date)
Wavebob	Wavebob	Surface – point absorber	Ireland	1:4 scale deployed in 2006

Key components

Buoys

Point absorber buoys tend to fall into one of two categories: a simple single-piece device and more complex two-piece buoys. Single-piece buoys move up and down with the motion of the wave, whereas the more complex two-piece buoys extract energy by a relative motion between its two parts. Some, but not all, two-piece buoys are sub-surface devices and are filled with air. As the peak (maximum height) of a wave passes over these buoys, the external water pressure above the buoy increases, forcing the buoy downwards, which creates a linear motion similar to that of the surface-piercing buoys.

Coupling system

There are two main types of coupling systems used in conjunction with point absorbers: mechanical linkages and hydraulic systems. Mechanical linkages in their simplest form consist of a cable which is coupled directly to a linear generator. In other more complicated mechanical linkage systems, rack-and-pinion gear systems are used to convert the linear motion of the buoy to a rotational motion. Devices such as the Aegir dynamo mechanical drive designed by Ocean Navitas Systems use a specially designed dual-gear system to convert the upward and downward motion of the buoy into singular directional rotational energy. Hydraulic systems use the linear motion of the buoy to produce a high-pressure working fluid which can then be fed into a hydropower turbine to create rotational energy. In some near-shore devices, high-pressure fluid is pumped to a hydropower turbine on land.

Generators

Point absorbers are typically coupled to either a linear or conventional generator, depending upon whether a system for converting the linear motion of the buoy to rotation motion is utilized. A generator consists of two main parts: the rotor and stator. In a conventional generator, the rotor is the rotating part and the stator is the non-moving part. In a linear generator, the motion of the 'rotor' is linear. In each instance, when the rotor moves relative to the stator, a magnetic field is created, causing current flow and the production of electricity.

Anchoring

Point absorbers can either be directly anchored to the seabed or float on the surface of the sea where the absorber is anchored to the seabed via a mooring line. Typically, floating types utilize the two-piece buoys.

Deployment location

Typically, point absorbers are located up to 12 nautical miles offshore in water depths of over 40m, although some near-shore devices have been proposed.

3.2.2 Attenuator

Working principle

Attenuator floating devices effectively 'ride' the waves and selectively constrain movement along their length to produce energy. The device is a long snake-like structure, which consists of a number of segments joined by hinges, anchored at one end to the sea floor. This arrangement enables the device to float on the surface of the sea, aligning itself head on to the oncoming waves. The force of oncoming waves causes the hinges to flex and the whole structure weaves along the surface. Hydraulic pistons at the hinges pressurize fluid that is used either within the device or pumped to an external generator to produce electricity. Active control of the stiffness alters the device's natural frequency to resonate with the given frequency of the waves and achieve a higher energy yield.

Attenuator devices have a relatively small area exposed to the face of the waves, enabling them to reduce the hydrodynamic forces of inertia, drag and slamming that have the potential to cause significant damage to offshore devices. Nevertheless, these devices feature a number of moving parts, including a complex hydraulic system comprising rams, accumulators and motors.

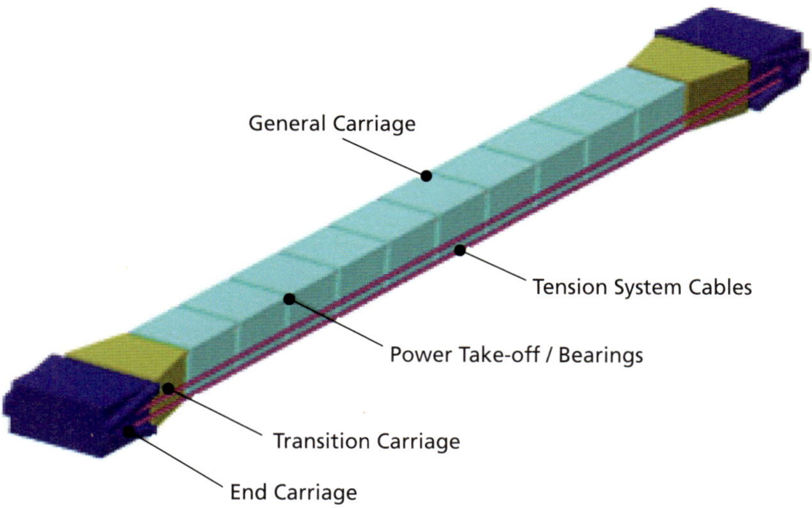

Figure 3.3 Wavegen HYDRA

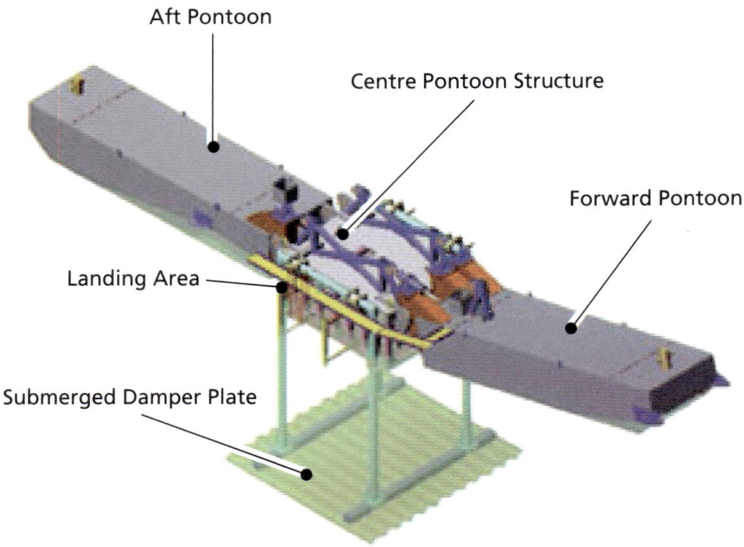

Figure 3.4 McCabe wave pump
Source: Polaski (2003)

Figure 3.5 Pelamis wave energy converter at sea
Source: Pelamis Wind Power

Available and researched designs

Table 3.3 summarizes some available and researched designs.

Table 3.3 Summary of past and existing designs of attenuator devices

Company	Technology	Device Type	Country Base	Stage of Development
Wavegen	HYDRA	Surface attenuator	UK	A development programme was cancelled in 2000 before prototype stage because calculations showed that the likely performance did not justify the construction and deployment of a prototype.
McCabe	Wave Pump	Surface attenuator	Ireland	A full-size 40m prototype was tested off the coast of Ireland in 1996, but no known commercial deployments to date (US Department of the Interior, 2006).
Pelamis Wave Power	Pelamis	Surface attenuator	UK	First trialled in Orkney 2004 and three commercial devices were installed off the Portuguese coast in 2008. There are plans for five further devices to be installed in British waters in the near future (OPD, 2006).

Key components

Energy take-off

In order to extract power from the motion of the device, hydraulic rams force fluid into accumulator vessels. Accumulator vessels enable smoothing of the power output by acting as reservoirs of high-pressure fluid. A hydraulic motor is used to generate electricity from the pressurized fluid.

Control system

In order to extract the maximum amount of power, it is necessary for the response of the attenuator device to be matched with the excitation over time.

When the stiffness of the joints is controlled correctly, it allows the natural frequency of the device to resonate with the wave frequency.

Fixings/moorings

Floating devices need to be held in position by a mooring system that comprises a combination of floats and weights in order that the mooring cables are prevented from becoming taut; this allows the device to swing into the oncoming waves.

Carriages

The carriages are designed to withstand the force of the waves, to be buoyant and to house some power generation or transmission equipment (depending upon the design). The design of the carriages currently appears arbitrary: Pelamis used a circular section; the HYDRA study concluded that rectangular and D-shaped sections were interchangeable.

Deployment location

Attenuator devices are typically designed to be moored in waters approximately 50 to 70m in depth (usually 5 to 10km from the shore) where the high energy found in deep swell waves can be accessed. The nature of the 'slack' mooring system requires the use of large sea areas to prevent devices from colliding.

The first and only commercial application to date comprises three 750kW Pelamis devices which began operation 5km off the northern coast of Portugal in September 2008 as part of project Aguçadoura. By November 2008, all three devices had to be towed back to harbour for mechanical repairs. It is believed that, due to commercial issues, the devices have not been re-launched to date.

E.ON has placed an order for a second-generation machine known as the 'P2' and in doing so has signed the UK's first commercial supply contract within the marine sector. It is believed that the second-generation device will be located at the same site in Orkney from which the Pelamis prototype was tested in 2004.

A 3MW (four units) development is planned in the Orkney Isles by ScottishPower Renewables, utilizing the existing electrical sub-sea cables, substation and grid connection used for the prototype trials in 2004. Funding has been announced by the Scottish government and the project has planning consent for construction, but has yet to be realized.

3.2.3 Oscillating water column

Working principle

Oscillating water columns (OWCs) consist of a partly submerged chamber with an inlet allowing seawater to flow in and out freely. As waves enter the chamber, the level of water rises, compressing the air in the top of the chamber, which in turn drives an air turbine. When the water inside the chamber recedes as the waves outside draw back, the air is sucked back under pressure into the chamber, keeping the turbine moving. The turbine is designed to rotate in the same direction irrespective of the direction of the airflow. The air turbine is connected to a generator to produce electricity.

Available and researched designs

Table 3.4 overleaf summarizes some available and researched designs of oscillating water column devices.

Key components

Chamber

The chamber is an air-filled space that is partly submerged in water. The size and number of chambers employed varies between devices, with the larger multi-chambered devices based onshore. Air flow is generated when the water level in the partly submerged chamber rises or falls due to passing waves. The actual air flow depends upon the wave head.

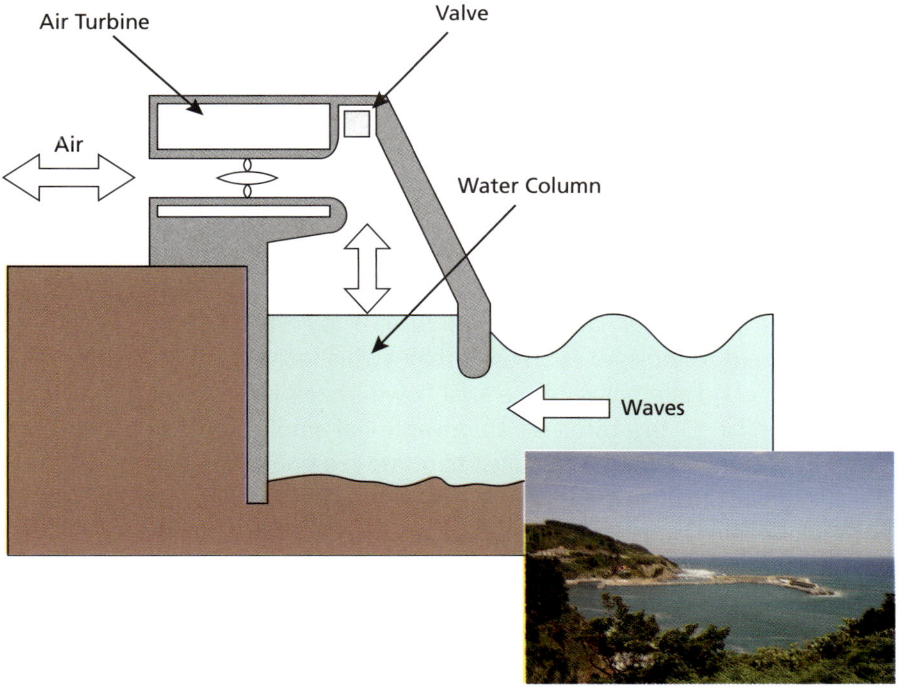

Figure 3.6 Oscillating water column schematic (left); Mutriku Wave Power Plant (right)
Source: left: Wavenet (2003), right: Voith Hydro

Table 3.4 Available designs of oscillating water columns

Company	Technology	Device Type	Country Base	Stage of Development
Ocean Energy Ltd	Ocean Energy Buoy	Oscillating water column	Ireland	Prototype tested
Oceanlinx (formerly Energetech)	Denniss-Auld Turbine	Oscillating water column	Australia	Joint venture between Iberdrola and Tecnalia – prototype tested
Voith Hydro Wavegen	Limpet/ Hywave	Oscillating water column	UK	Limpet was installed in 2000 and continuously updated with new turbines; the technology has been installed in breakwater in the Mutriku Wave Power Plant (Spain) – the first worldwide commercial operation, inaugurated in July 2011

Air turbine

Several types of air turbines have been considered; however, bi-directional flow turbines are viewed more favourably as they take advantage of the air flow in both directions. Generators produce power by spinning in one direction; therefore, having a turbine which rotates in the same direction regardless of the direction of the air flow driving it is very advantageous. To achieve this, air turbines have specially shaped blades. The Wells turbine developed during the 1970s and the newer Denniss-Auld turbine by the Australian company Energetech are examples of axial flow turbines which achieve the aforementioned characteristic. An alternative to the axial flow turbines is the impulse turbine; its rotor is similar to the rotor of a conventional single-stage steam turbine. However, since the turbine is required to rotate in a single direction regardless of the direction of the air flow, two rows of guide vanes, placed symmetrically on both sides of the rotor, are used instead of the conventional single row.

Generator

Oscillating water columns use conventional generators to convert the kinetic energy of the air turbines to electrical energy.

Deployment location

Devices such as the Limpet are located onshore; however, the technology can also be integrated within newly built or retrofitted breakwaters. Moreover, near-shore applications with purpose-built walls, as in the 4MW Siadar Wave Energy Project, are possible. Here the walls stand in water depths of approximately 10m and several 100m offshore in the surf zone.

3.2.4 Oscillating wave surge collectors

Working principle

Oscillating wave surge collectors operate by capturing the wave energy directly, without the need for any collector. Waves push against a flat/float/paddle, which is hinged against a fixed reaction point. The paddle is connected to a pump, fitted with double-acting water pistons. As the waves push the paddle backwards, the hydraulic piston creates a high-pressure fluid which converts to electrical power using conventional hydroelectric generators.

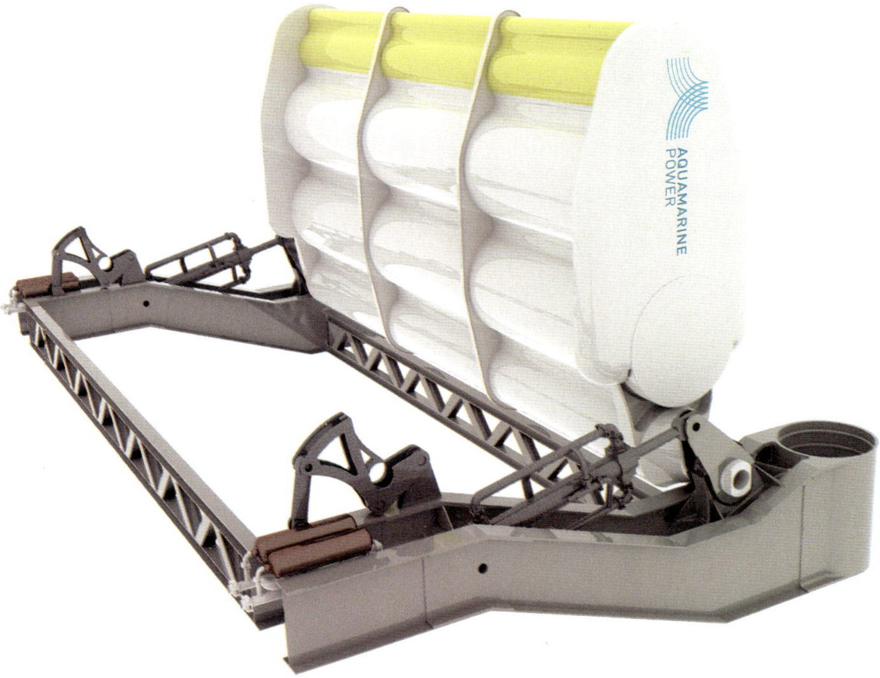

Figure 3.7 Oyster
Source: Aquamarine Power www.aquamarinepower.com

Available and researched designs

Table 3.5 summarizes some available and researched designs of oscillating wave surge devices.

Table 3.5 Available oscillating wave surge designs

Company	Technology	Device Type	Country Base	Stage of Development
Aquamarine Power	Oyster	Oscillating wave surge converter	UK	November 2009: first demonstration-scale (connected to national grid – 315kW) The joint venture partnership is currently developing a 2MW demonstration site for 2011, to be expanded to 10MW in 2012 and 200MW thereafter
AW Energy	Waveroller	Oscillating wave surge converter	Finland	October 2009: signed a €3 million contract with the EU to demonstrate its technology
BioPower Systems Pty Ltd	bioWave	Oscillating wave surge converter	Australia	Two 250kW pilot projects scheduled to be operational during 2010
Floating Power Plant ApS (FPP)	Poseidon's Organ	Oscillating wave surge converter	Denmark	The 37m wide, 350 tonnes demonstration plant was launched in Nakskov Harbour in the summer of 2008
SDE	SDE	Oscillating wave surge converter	Israel	40kW unit tested

Key components

Paddle/float

The paddles/floats can vary in design depending upon where the hinge point is located. Paddles can be hinged at the top of the device if the paddles are above sea level or at the bottom of the device if the system is anchored to the seabed. The paddles must be able to withstand large forces and are typically very substantial, rigid structures.

Hydraulic system

The paddle is hinged around a fixed reaction point which is coupled to large hydraulic pistons, which comprise two parts: a ram and a cylinder. When these rams are pushed backwards, a high-pressure working fluid is created in the cylinder. This high-pressure fluid is fed into a hydraulic turbine (typically, an impulse turbine, such as a Pelton wheel), which creates a rotational energy, which in turn is used to drive a generator.

Generator

Oscillating wave surge collectors use conventional generators to convert the kinetic energy of the hydropower turbine to electrical energy.

Deployment location

Oscillating wave surge collectors are shoreline or near shoreline wave converters. The subsurface devices are typically at a depth of 8m to 10m.

3.2.5 Overtopping

Working principle

Overtopping devices elevate ocean waves into a reservoir above sea level. The water in the reservoir creates a head, with the difference between the 'normal' level of the water surface and the water surface in the reservoir acting like a dam. Water is returned to the sea through a number of turbines generating electricity.

Available and researched designs

Table 3.6 summarizes some available and researched designs of oscillating wave surge devices.

Figure 3.8 Wave Dragon
Source: Wave Dragon www.wavedragon.net © Earth-vision.biz

Table 3.6 Overtopping designs

Company	Technology	Device Type	Country Base	Stage of Development
Wave Dragon	Wave Dragon	Overtopping device	Wales/ Denmark	7MW unit planned for development and deployment
Wave Energy	Seawave Slot-Cone Generator	Overtopping device	Norway	200kW unit being constructed
WavePlane Production	Wave Plane	Overtopping device	Denmark	First full-scale prototype deployed (large)

Key components

Reservoir, ramps and reflectors

Overtopping devices can be designed to be shoreline or offshore devices. Shoreline devices and those with foundations tend to be larger structural devices and some designs have multiple reservoir chambers. Offshore overtopping floating devices are slack moored, similar to systems used for ships.

Ramps can be compared to a beach. When the wave reaches a beach it encounters a new surface which exerts different new forces and pressures on the wave. The forces and pressures are strong enough to change the nature of the wave, allowing all the energy to be dissipated as the wave breaks at the shoreline. Ramps are designed to change the wave's geometry and to elevate it. They are specially shaped to optimize this effect.

Reflectors are used to direct waves into the reservoir. The Wave Dragon, an offshore device, uses double reflectors to deflect the waves into the reservoir. As the waves reach the reflectors, they elevate and reflect towards the ramp, increasing the amount of overtopping water and thereby increasing the possible energy output.

Propeller turbines

The turbines used in overtopping devices are the same as those used in traditional hydroelectric power plants. Kaplan turbines work on the basis that the water changes pressure as it moves through the turbine, giving up its energy. The water is directed on to a propeller-shaped runner, causing it to spin.

These turbines are designed to be run at full speed as the efficiency tends to drop off quickly. In some applications, multiple turbines are used, with each turbine regulated by starting and stopping turbines individually using water control gates. This controls the total water outlet and power production efficiently.

Generator

Overtopping devices use conventional generators coupled to the same shaft as the propeller turbines, converting the kinetic energy of the turbine to electrical energy.

Deployment location

Overtopping devices can be shoreline or offshore wave converters. The offshore devices are typically operated at depths of 25m to 40m.

3.2.6 Costs

The Carbon Trust estimates the capital cost for first prototype and first production wave energy converters to be up to €10,350/kW (UK£ to Euro exchange rate of 1.15), but did note that certain prototypes have been built for costs below €4945/kW.[6] Further to this, the capital costs of wave installation were predicted to be in the range of €1955/kW to €4945/kW. The UK Department of Trade and Industry (DTI) placed the capital cost in the range of €1955/kW to €4600/kW,[7] and also highlighted that the cost of near or shoreline wave converters will be greater than offshore designs.

It should be noted that while it is possible to extrapolate from the cost of a prototype to a commercial installation, the figures presented appear to be optimistic for first commercial generation installations. However, like all new technologies, as experience is gained and the industry grows, the capital cost is expected to reduce.

Operating costs for wave installations are estimated between €13.8 cents/kWh and €50.6 cents/kWh (Carbon Trust, 2006). The wide range is indicative of the large variation in technologies available, as well as substantial uncertainty about capacity factor, performance and operating and maintenance costs.

Certainty regarding costs for wave projects will only be achieved once a number of projects have been realized.

3.3 Tidal energy devices

This section provides a brief overview of the tidal energy technologies that could harness the tidal resource. The following technologies are considered:

- horizontal axis turbine;
- vertical axis turbine; and
- oscillating hydrofoil.

Many designs are currently being developed, including novel approaches such as those that use vortex-induced vibrations to generate electricity, as well as the technologies considered for this section. Several of the turbine devices use uni-directional turbines rather than bi-directional turbines. Currently, no design or approach is considered to be a leading design or best solution.

Towards the end of this section, alternative technologies, such as osmotic and ocean thermal, are briefly discussed.

3.3.1 Horizontal axis turbine

Working principle

Horizontal axis turbines use the aerodynamic profiles of their turbine blades to convert the kinetic energy of the flowing water into rotational motion, which in turn is converted into electrical energy using a generator. Horizontal axis turbines have their axis of rotation parallel to the motion of the water. Horizontal axis turbines are largely based on wind turbine designs; however, the majority have bi-directional blades, allowing the turbine to spin with both tidal directions. Other designs are mounted on a crossbeam which swings 180 degrees as the tide changes direction. Designs have been developed to operate with or without a gearbox.

Open-centre generators have a different design where the rotor upon which the blades are mounted acts as the rotor to the generator, with no central shaft.

Figure 3.9 Tidal stream turbine
Source: Hammerfest-strøm

Figure 3.10 HyTide 110/5.3
Source: Voith Hydro

Figure 3.11 Open-centre turbine
Source: OpenHydro

Available and researched designs

Table 3.7 summarizes the available and researched designs of horizontal axis tidal devices.

Key components

Blades

The blades of horizontal axis turbines have a specific aerodynamic profile which enables them to convert the kinetic energy stored in the water into torque. The number of blades varies with the design, although typically, as in wind turbines, two- or three-bladed designs are common. Unlike wind turbines, some tidal turbines have blades which are specially shaped to allow them to rotate in both directions, enabling them to operate both when the tide is coming in and going out.

Nacelle

The nacelle, such as in wind turbines, is the 'house' of the main components of the tidal turbine. In designs where the nacelle is submerged, the nacelle is pressurized with an inert gas to avoid water ingress. Not all tidal turbines incorporate a nacelle.

Gearboxes

Gearboxes are used in some instances to increase the speed of the shaft to which the blades are attached. This converts low-speed high-torque rotational

Table 3.7 Horizontal axis tidal designs

Company	Technology	Device Type	Country Base	Stage of Development
Hammerfest Strom	Tidal Stream Turbine	Horizontal axis turbines	Norway	300kW machines prototype tested developing 1MW prototype for commercial application
Marine Current Turbines	Seagen, Seaflow	Horizontal axis turbines	UK	1.2MW unit operational
Ocean Flow Energy	Evopod	Horizontal axis turbines	UK	One-fifth scale (22kW) prototype operational
OpenHydro	Open Centre Turbine	Horizontal axis turbines	Ireland	1MW commercial turbine deployed
Hydratidal	Tidal turbine	Horizontal axis turbines	Norway	First unit under construction due to be launched in 2010
Voith Hydro Ocean Current Technology	HyTide 110/5.3	Horizontal axis turbines	Germany	110kW turbine operating in Korea since 2011
Atlantis Resource Corporation	AK-1000	Horizontal axis turbines	Singapore	Full-scale prototype

energy into high-speed low-torque rotational energy, which can be passed on to a conventional type generator.

Direct-drive generators

Direct-drive generators eliminate the need for gearboxes. Conventional generators rotate at a high speed, around 2000 rpm; however, direct-drive generators operate at the same rotational speed as the turbines, which for tidal applications is much slower. Generally, permanent magnet generators are the most suitable for direct-drive applications, with a permanent magnet rotor and wound stator.

Open-centre turbines

The open-centre turbine is a specific type of tidal turbine which varies significantly from the more traditional turbine design. It consists of a single-piece rotor, with blades mounted on an inner, open-centred hub and the outer edge of the rotor, housed concentrically within the stator of the generator. The outer edge of the blades acts as part of the generator. The rim generator operates at the same rotational speed as the turbine, making it a direct-drive permanent magnet machine.

Deployment location

The limiting factors are tidal speed, depth of water and whether foundations can be easily established.

3.3.2 Vertical axis turbine

Working principle

Vertical axis turbines use the aerodynamic profiles of their turbine blades to convert the kinetic energy of the flowing water into rotational motion, which in turn is converted to electrical energy using a generator. Vertical axis turbines have their axis of rotation perpendicular to the motion of the water and consist of a number of evenly spaced blades mounted on support arms, arranged concentrically to a central rotational shaft (rotor). Water motion drives the blades, which in turn rotate the central shaft. The central shaft is geared and drives a conventional generator to produce electricity. Vertical axis turbines usually have the rotor mounted into a concrete plinth, which anchors the unit to the ocean, with the power take-off equipment above the water surface.

The majority of vertical axis turbines tend to be included in tidal barrage structures; therefore, the offshore applications are more limited.

Available and researched designs

Table 3.8 summarizes some available and researched designs of vertical axis tidal devices.

Table 3.8 Vertical axis tidal designs

Company	Technology	Device Type	Country Base	Stage of Development
New Energy Corp	EnCurrent Vertical Axis Hydro Turbine	Vertical axis turbines	Canada	25kW station deployed, 125kW and 250kW machines due 2010
Blue Energy	Davis Turbine	Vertical axis turbines	Canada	Raising money for a 200MW project

Figure 3.12 EnCurrent vertical axis hydro turbine
Source: New Energy Corporation 2010 http://www.newenergycorp.ca/

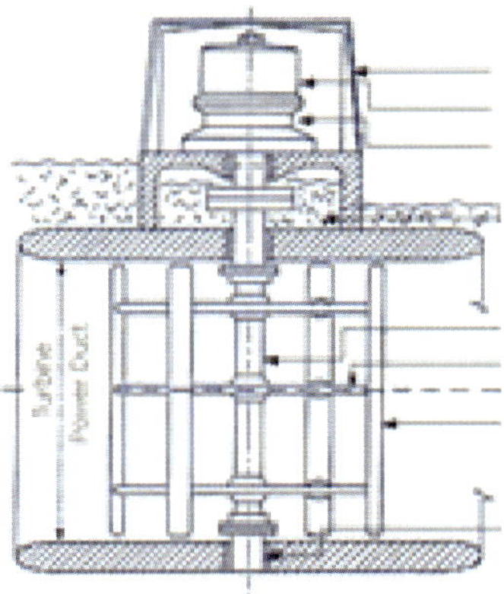

Figure 3.13 Blue Energy Davis turbine
Source: http://www.bluenergy.com/Technology.htm

Key components

Blades

Vertical axis turbines employ evenly spaced blades coupled to a central shaft. Each blade consists of a symmetrical hydrofoil cross-section. The blades are specially designed to allow them to move proportionately faster than the motion of the water. The symmetrical hydrofoil design means that the turbine can rotate in the same direction with water flow from either direction.

Gearbox

A gearbox is used to increase the speed of the shaft to which the blades are attached. This converts low-speed high-torque rotational energy into high-speed low-torque rotational energy which can be passed on to a generator.

Generator

Vertical axis turbines currently use conventional generators coupled to the output shaft from the gearbox; this converts the kinetic energy of the shaft to electrical energy.

Deployment location

The limiting factors are tidal speed, depth of water and whether foundations can be easily established. Vertical axis turbines can be placed in shallower waters or in narrower channels than horizontal axis turbines.

3.3.3 Oscillating hydrofoil

Working principle

Oscillating hydrofoil devices are seabed mounted, consisting of one or more hydroplanes with variable angle of attack to which a pair of pivoting arms is attached. The action of the tidal stream over the hydroplane causes the arms to lift and fall; when the arms reach the top of their stroke the hydroplane incline angle is reversed, causing the arm to be lowered and to start the cycle again. The arms are linked to hydraulic rams which pump hydraulic oil to accumulators powering hydraulic motors that drive a generator.

Available and researched designs

Table 3.9 summarizes some available and researched designs of oscillating hydrofoil devices.

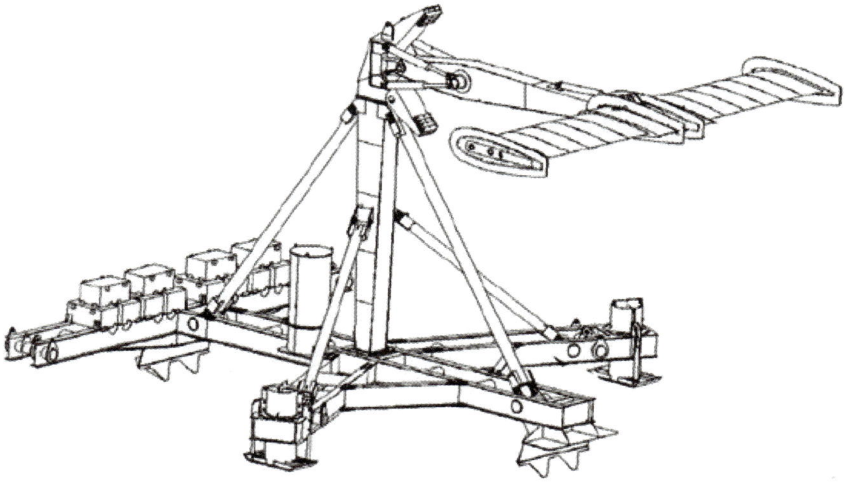

Figure 3.14 Stingray
Source: Instituto Superior Técnico, Technical University of Lisbon, Portugal

Table 3.9 Oscillating hydrofoil designs

Company	Technology	Device Type	Country Base	Stage of Development
Pulse Tidal	Pulse Stream	Oscillating hydrofoil	UK	100kW prototype tested
The Engineering Business	Stingray	Oscillating hydrofoil	UK	150kW prototype tested

Key components

Hydrofoil

The hydrofoil is a wing-like structure which, depending upon the angle of attack, is positive or negative to the tidal stream in-flow; the hydrofoil will rise and fall in an oscillating motion. The hydrofoil is mechanically linked to a support structure and pivots around a fixed point.

Hydraulic system

As the hydrofoil pivots around the fixed reaction point it actuates a number of hydraulic pistons. The pistons comprise two parts: a ram and a cylinder. When these rams are pushed backwards, a high-pressure working fluid is created in the cylinder. This high-pressure fluid is fed into a hydraulic turbine, creating a torque. The hydraulic turbine is coupled to a generator which, in turn, produces electricity.

Control system

The angle of attack of the hydrofoil must be controlled and dynamically optimized in order to maintain efficient performance. This is achieved by programme logic control (PLC), which controls the hydraulic rams mounted on the oscillating arm in order to actuate the hydrofoil angle of attack. This system is used to overcome the pitching moment on the hydrofoil, stop its motion and return the cycle in the opposite direction.

Generator

Oscillating hydrofoils use conventional generators coupled to the output shaft from the hydraulic turbine; this converts the kinetic energy of the shaft to electrical energy.

Deployment location

This device is a seabed-mounted machine, to be situated typically in any water depth of up to 100m.

3.3.4 Add-on devices: Venturi

Working principle

Venturis create a vortex of low pressure behind a turbine, which increases the flow over the turbine and allows the turbine to operate at higher efficiencies. The flow of water can either drive a turbine directly or an air-turbine (taking advantage of the induced pressure differential in the system).

Available and researched design

Table 3.10 summarizes some available and researched designs of venturi add-on designs.

Table 3.10 Venturi add-on devices

Company	Technology	Device Type	Country Base	Stage of Development
Hydro Green Energy	Hydrokinetic Turbine	Venturi Effect	US	100kW unit tested
Tidal Energy Pty Ltd	DHV Turbine	Venturi Effect	Australia	Prototype tested

Key components

When fluid enters the constricted section of pipe in a venturi, the fluid velocity increases as the pressure drops. For tidal applications, the widest section faces downstream and the narrow opening faces into the flow of water, as this produces the greatest increase in the turbine performance. The downstream venturi produces a sub-atmospheric low pressure behind the inlet, augmenting the flow and increasing the flow twofold to fourfold. A turbine, either axial or cross-flow, located here operates at higher efficiencies. Additional gaps in the venturi can inject flow, from outside, helping to maintain a boundary layer, reinvigorating the flow behind the turbine and further improving efficiency.

Deployment location

Venturis are add-on devices improving the efficiency of tidal devices; therefore, their deployment is directly linked to the tidal devices with which they can be used.

3.3.5 Costs

The Carbon Trust commissioned a report investigating the capital costs of first prototype and first productions of tidal energy converters, concluding the price range to be up to €9200/kW, but did note that certain prototypes have been built for costs below €5520/kW.[8] Further to this, the capital costs of tidal installation were predicted to be in the range of €1610/kW to €3450/kW. The DTI placed the capital cost in the range of €2300/kW to €4600/kW.[9]

It should be noted that similar to wave projects, while it is possible to extrapolate from the cost of a prototype to a commercial installation, the figures presented appear to be optimistic for first commercial-generation installations. However, like all new technologies, as experience is gained and the industry grows, the capital cost is expected to reduce.

Operating costs for tidal installations are estimated between €10.35 cents/kWh and €20.7 cents/kWh (DTI, 2007). The reason for the smaller range compared with wave devices is twofold: less variation in the prototype designs being proposed and more confidence in capacity-factor predictions. It should, however, be noted that potential operation and maintenance costs are still uncertain.

Certainty regarding costs for tidal projects will only be achieved once a number of projects have been realized.

3.3.6 Alternative designs

The following technologies, although outside the scope of this book, are briefly presented for completeness and as an introduction to their potential.

Ocean thermal energy converters

Ocean thermal energy converters (OTECs) are heat engines which produce energy due to the temperature difference that exists between deep and shallow waters. A heat engine is a thermodynamic device placed between a high temperature reservoir and a low temperature reservoir. As heat flows from one to the other, the engine converts some of the thermal energy into kinetic energy. The system is based on the Rankine cycle, using a low-pressure turbine. The greatest efficiency and power is produced at the largest temperature differences.

Lockheed Martin's Alternative Energy Development team is currently in the final design phase of a 10MW OTEC system, which is due to become operational in Hawaii in 2012 to 2013.[10]

Osmotic power generation

Osmosis occurs wherever two solutions of different concentrations meet at a semi-permeable membrane. Two methods of power generation are available: pressure-retarded osmosis and reverse electrodialysis. Salt and freshwater are the two solutions used.[11]

Pressure-retarded osmosis works with the spontaneous passage of water from dilute to concentrated solutions through a membrane. This generates a pressure difference that can be harnessed to generate power. The two liquids, salt and freshwater, are pumped to either side of a membrane, where osmosis creates a pressure equivalent to a column of water over 100m high (11 to 15 bar being the optimum pressure). This is used to drive a turbine and generate electricity.

The Tofte plant in Norway is currently producing 4kW of energy.[12]

In reverse electrodialysis, a salt solution and freshwater passes through a stack of alternating cathode and anode exchange membranes. The chemical potential difference between salt and freshwater generates a voltage over each membrane.[13]

A 50kW research plant is located at Harlingen, the Netherlands.

3.4 Foundations, moorings and grid connection

This section provides a brief overview of the remaining offshore components of a project – namely, the foundations and electrical connection to the grid.

3.4.1 Foundations

Foundations can be broken down into six broad categories: monopile, 'multipile' (tripile and tripod), gravity base, jacket, suction cup and floating foundations. Table 3.11 summarizes the various foundation types and the anticipated water depth limits. While each of the aforementioned designs are valid in their own right, only three are expected to be suitable for deeper water (typically over 20m for wind farms currently under construction[14]) and non-homogeneous soil conditions: 'multipile', jacket and floating foundation designs.

Table 3.11 Comparison of foundation types

Foundation Type	Water Depths <20m	Water Depths >20m	Complex Soil Conditions	Experienced?
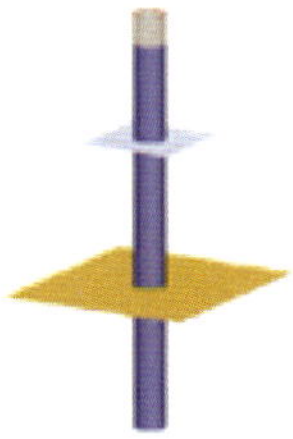 Monopile Foundation	Yes	(Up to around 25m)	Unsuitable for stony and hard soils	Yes
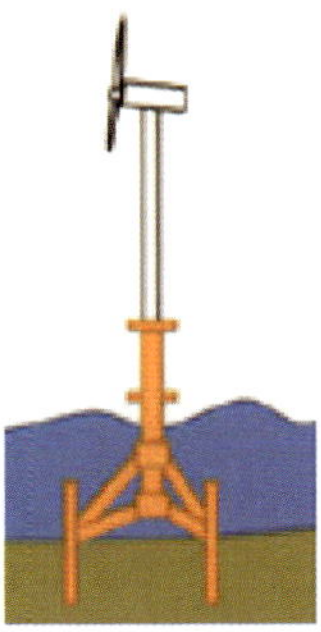 Tripod Foundation	Limited due to more economical solutions	Yes	Unproven; however, small diameter of piles would enable drilling of pile through stony/hard soils	Limited
 Tripile Foundation	Limited due to more economical solutions	Yes	Based on design (3 monopiles) not likely to be suitable	Limited

Table 3.11 continued

Foundation Type	Water Depths <20m	Water Depths >20m	Complex Soil Conditions	Experienced?
 Jacket Foundation	Limited due to more economical solutions	Yes	Yes, due to small pile size	Yes
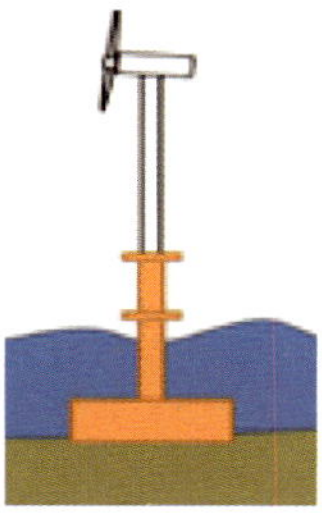 Gravity Foundation	Yes, with surface preparation	No	Soil needs to have sufficient bearing capacity and will most likely require preparation	Yes
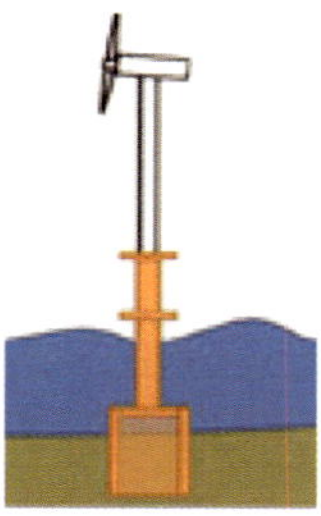 Suction Cup Foundation	Theoretically	No	Requires soft and homogeneous soils to function	Limited
 Floating Foundation	No	Yes, for very deep water	Yes, piles not necessary	No; however, currently under test in the HyWind project

Source: Deutsche WindGuard

Even though the foundations illustrated above have been considered for offshore wind projects, it is anticipated that similar designs will be used for tidal projects.

Multipile

The group of 'multipile' is characterized by a steel structure with more than one (typically three) penetrating the soil on the seabed. In principle, there are two competing designs. The first is the Tri-Pile designed by BARD, which uses three monopiles joined above sea level to form the foundation. In comparison to this, WeserWind has designed a 'Tri-pod' design which uses a steel structure joining the three load-carrying piles under the sea surface and with one central column. Key advantages of both designs are similar in so far that they are suitable for deep water and inhomogeneous soil conditions with stony subsoil due to the relatively small diameter of each of the piles used to secure the foundations to the soil.

Lack of practical experience with either system, manufacturing complexity and relative mass are all considered drawbacks. The weight of these foundations will require special lifting equipment that is currently scarce. However, according to Windenergie Agentur Bremen,[15] the shipyards (namely, MPI and GustoMSC) have anticipated this trend and currently have plans to deliver 12 new vessels capable of handling these types of foundations.

Jacket

Jacket foundations are well known from the offshore oil and gas industry. The design is fairly straightforward and changes in water depth require little change in design. The specific weight of the foundation compared to other solutions is also low. However, the construction is labour intensive and requires large manufacturing capacity adjacent to suitable port facilities. The fact that the design is relatively easy to construct, however, widens the availability of production facilities, making this type of foundation available on a large scale.

Floating foundations

Floating foundations are also well known from the offshore oil and gas industry. Prototypes have been installed and the results are expected with anticipation from the industry as this could result in making far offshore locations more attractive for development, as well as lead to reduced costs.

The prime concept of a floating foundation is a counterweight under the sea surface in the shape of a tube or other structure. The counterweight provides sufficient weight to lower the centre of gravity of the entire structure, thus keeping the structure upright and stable while at the same time providing buoyancy to the entire structure. Floating foundations are fixed in position by the use of guy wires or a similar method. The export cable hangs off the structure and is fixed to the seafloor. Currently, only limited experience of floating foundations is available to the wind industry.

3.4.2 Moorings

Some marine energy devices will require moorings. The main role of a mooring system is to allow the floating device to sustain its position, otherwise known as 'station keeping'.[16] The mooring system can be used in two ways: either purely as a structural component keeping the device stable or by utilizing energy if the mooring system is part of the power train. Mooring system design, therefore, can vary depending upon the marine floating device design. The mooring system must also have a lifetime similar to that of the associated marine energy device. Table 3.12 describes different mooring configurations and their suitability for marine energy devices. References provided within this table can be consulted for a more detailed description of the various configurations presented.

Table 3.12 Comparison of different mooring types

Mooring Type	Characteristics	Cost
Spread Mooring		
Single Catenary 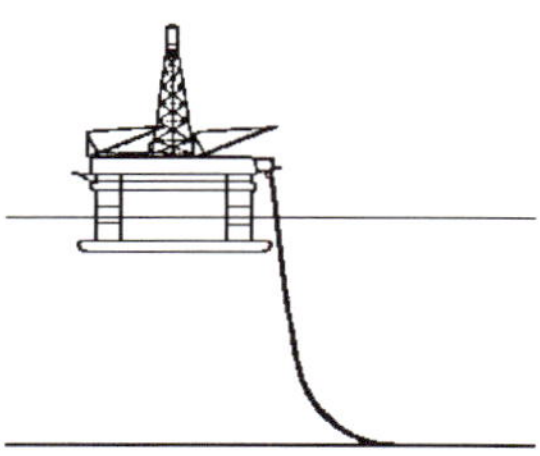	The mooring line is free hanging and horizontal to the seabed.	Low
Multi Catenary 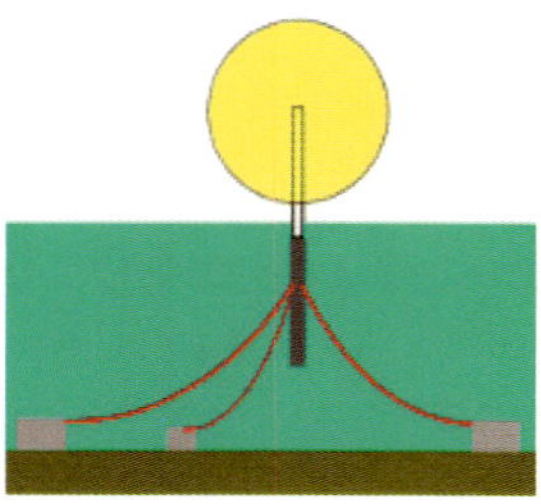	This mooring is the same as the single catenary but in multiple forms.	Low
Taut spread/tethered 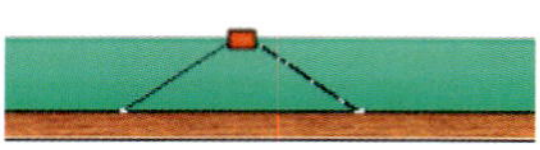	This mooring line is tensioned and is at an angle to the seabed.	Medium

Mooring Type	Characteristics	Cost
Single-point mooring		
Turret mooring 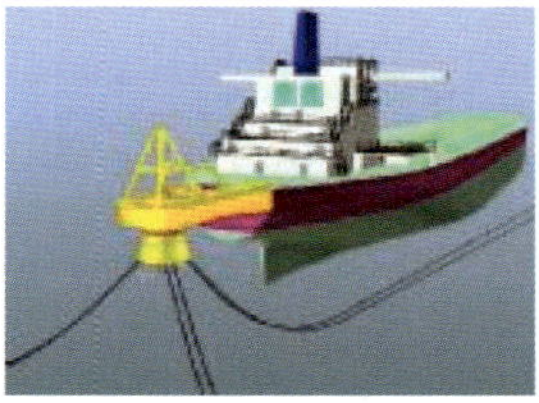	This mooring line consists of an internal/external catenary moored to a turret which is attached to the floating structure.	High External cheaper than internal
Catenary anchor leg (CALM) 	The floating device is moored to a buoy which is moored to a catenary moored anchor.	Low
Single anchor leg (SALM) 	The floating device is moored to a buoy which is moored to a taut moored anchor.	Low
Articulated loading column (ALC) 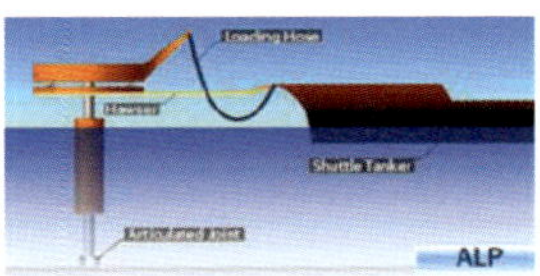	The floating device is moored to a bottom hinged column.	Medium
Single point + reservoir (SPAR) 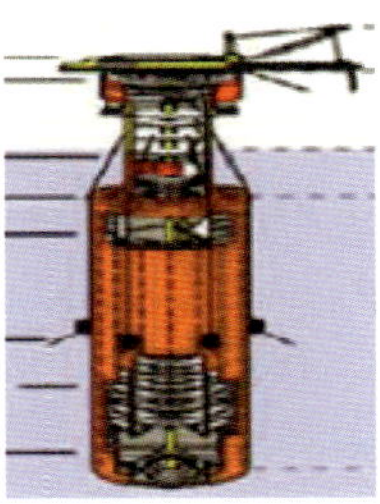	The floating device is catenary moored, but as part of the mooring system an energy store is provided.	Medium

Table 3.12 continued

Mooring Type	Characteristics	Cost
Single-point mooring (cont.)		
Fixed tower mooring 	The floating device is anchored to a fixed tower.	Medium
Dynamic positioning		
Active mooring	The mooring lines are spread around the float and one end of the mooring is held by a servo-controlled winch.	High
Propulsion  	The floating device is positioned in a fixed position above the seabed using propellers and/or thrusters that are computer controlled.	High

Mooring lines and anchor types

As part of the mooring system, the mooring line and anchor can also differ. Table 3.13 describes the different mooring line and anchor options that may be used with marine energy technology. References provided within this table can be consulted for a more detailed description.

Arrays

Ultimately, wave energy devices will eventually be deployed as wave farms or arrays. These arrays will also require mooring and anchor systems. Potential array designs are shown in Figure 3.15. As no devices have reached this stage yet, there is very little research and development into mooring array designs to connect each device.

Table 3.13 Comparison of mooring lines and anchor types

Mooring Type	Characteristics	Cost
Spread Mooring		
Chain 	There are specific grade strengths that can be used offshore; chains are suitable long term, but require regular inspection.	Medium
Wire rope 	Wire rope is made from strands. This wire rope is suitable for long-term mooring, especially tension mooring due to its elasticity. Extreme bending has to be avoided.	Low
Synthetic rope 	Several types of synthetic material such as nylon can be used for this rope. Axial compression, heating and fish bites are all problem areas.	High
Anchor		
Drag-embedded anchor 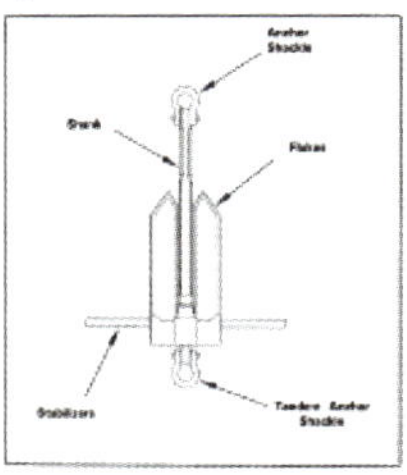	Due to the instalment direction of the anchor embedment, the horizontal load capacity is only in this direction.	Medium
Driven pile/suction anchor 	This anchor can take on horizontal and vertical loadings. This is due to the mechanical and pressure difference that was used to force the pile into the ground and the friction between the pile and the ground.	High

Table 3.13 continued

Mooring Type	Characteristics	Cost
Anchor (cont.)		
Vertical-load anchor 	This anchor can also take on horizontal and vertical loadings specific to the embedded anchor. Loads do not have to be in the same direction as the main instalment direction.	High
Drilled and grouted anchor	This anchor also has a horizontal and vertical load holding capacity, generated by grouting a pile in a rock that is a predrilled hole.	High

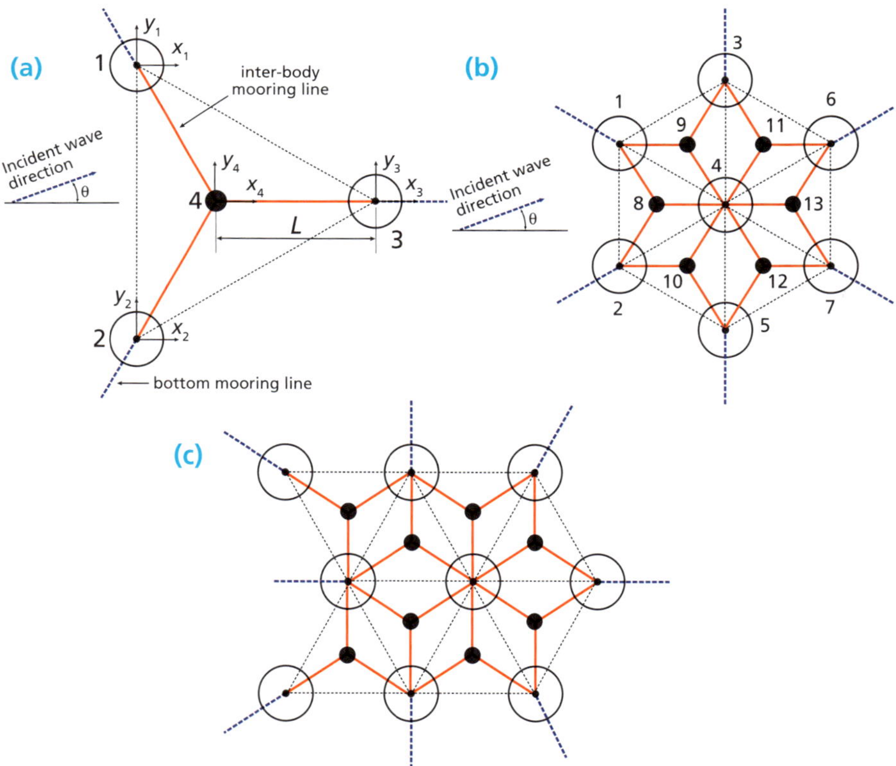

Figure 3.15 Potential mooring array designs

3.4.3 Electrical connections to shore

Grid connection offshore is typically a major undertaking. Figure 3.16 shows a typical setup of an offshore wind farm connection to shore.

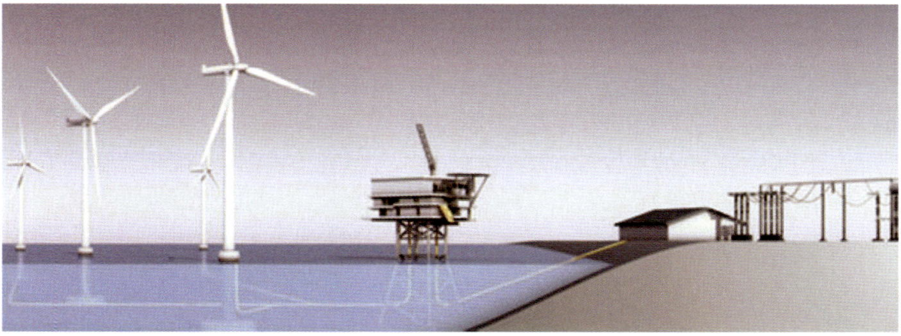

Figure 3.16 Typical wind farm grid connection arrangement

As can be seen in Figure 3.16, each turbine is connected to every other (via inter-array cables) and then to the offshore substation before it is connected to an onshore substation via an export cable. The installation process requires specialized marine equipment for the installation of the substation itself, as well as the inter-array cables (typically 33kV) and export cable (typically around 150kV).

Technically, the key issue is the loss in the transmission line associated with the line length. High voltage alternative current (HVAC) systems become less attractive due to transmission losses over a long distance and the use of high voltage direct current (HVDC) systems, which are in existence, will become more commonplace for far offshore projects. As a rule of thumb, HVAC systems are cost effective for cable lengths of 50km to 100km; above this length, HVDC systems are more cost effective and often the only practical solution. It should be noted that while HVDC is not new, there is limited experience offshore.

Soil conditions have an impact upon the installation of the grid connection. The cables are usually jetted into the ground; however, stony or hard soils necessitate the use of concrete mattresses or similar systems to fix the cable in its location.

A more detailed discussion of the technical challenges of electrical connection to shores appears in Chapter 8 (section 8.7).

Notes

1 Upwind: www.upwind.eu.
2 EWEA (2009b).
3 PFI Market Intelligence (2009).
4 EWEA (2009a).
5 PFI Market Intelligence (2009).

6 Carbon Trust (2006).
7 DTI (2007).
8 Carbon Trust (2006).
9 DTI (2007).
10 Lockheed Martin (2010).
11 Leonardo Energy (2010).
12 Webb (2009).
13 Leonardo Energy (2010).
14 EWEA (2010a).
15 WAB (2009).
16 Harris et al (2004).

4

Deployment Targets, Policies and Progress

4.1 Deployment targets and policies

This chapter provides a review of policies supporting the development of offshore renewables, discussing projects, device development and other supply chain capacity. The review provides an overview of recent policy implementation and programmes in the 18 countries of interest. It includes, where information has been readily available, a summary of the following aspects:

- processes for permitting, licensing and allocating seabed rights;
- support mechanisms for project deployment, including *price-based instruments* (e.g. concessional tariffs, carbon market mechanisms), *quantity-based instruments* (e.g. tendering systems, renewable portfolio standards, tradable certificates), *investment subsidies* (e.g. direct project capital subsidies, state-sponsored provision of new electrical interconnections) and *complementary measures* such as tax measures (e.g. production and investment tax credits, accelerated capital cost allowances, tax exemptions and rebates);
- support mechanisms for development of devices (e.g. governmental funding for research, demonstration projects and onshore or offshore testing facilities);
- support mechanisms for development of other supply chain capacity (e.g. industry informational resources and trade promotion activities).

Not all of the above information was readily available for every country. The analysis of support policies in section 9.15 (Chapter 9) complements the above review, filling in any gaps where possible and highlighting both successes and areas for development. It will also examine any trends in how policies vary according to the different legal status and technical challenges with increasing distance from shore (e.g. shoreline, within the 12 nautical mile zone, outside of the 12 nautical mile zone).

4.1.1 Support mechanisms

Nearly all of the 18 countries have some kind of policy support system in place for renewable energy generation, in general, with many specifying differentiated tariffs for offshore wind and marine energy. Table 4.1 highlights examples of such support measures among the 18 countries of interest.

Details of the national policies supporting the development of offshore renewables for the illustrated countries can be found in Appendix E ('Country profiles').

Table 4.1 Examples of offshore energy project support measures by country

Country	Key Project Support Measures	Price-based Incentive (€ c/ kWh)	Other Incentives
Canada	Federal incentive payment	All: €0.64 cents/kWh (+ wholesale tariff) for 10 years	Provincial RFPs, RPSs, and feed-in tariffs (FIT)
Denmark	Subsidy above market price	Wind: €3.7 cents/kWh (+ wholesale tariff) for ~8 years Marine: €8 cents/kWh (total) for 10 years €5 cents/kWh (total) for further 10 years	Individually tendered FITs for offshore wind sites e.g. €8.4 cents/ kWh
France	National FIT	Offshore wind: €13 cents/ kWh for 10 years €3–13 cents/kWh for further 10 years Marine: €15 cents/kWh for 20 years	
Germany	National FIT	Offshore wind: €15 cents/ kWh for 12 years €3.5 cents/kWh for further 10 years	
Ireland	National FIT	Offshore wind: €14 cents/ kWh for 15 years Marine: €22 cents/kWh for 15 years	
Italy	Green Certificates (>1MW) National FIT (<1MW)	Total tariff (green certificates + wholesale tariff): ~€18 cents/kWh for 15 years Marine FIT: €34 cents/kWh for 15 years	Tax incentives in southern Italy

Country	Key Project Support Measures	Price-based Incentive (€ c/kWh)	Other Incentives
Japan	Renewable portfolio standard		
The Netherlands	Federal incentive payment	Offshore wind: €18.6 cents/kWh for 15 years	Tax incentives
Norway	Capital grants		
UK*	National green certificates	Marine/offshore wind: ~€11 cents/kWh (+ wholesale tariff) for 20 years	Climate change levy exemption Capital grants
Belgium	Regional green certificates National minimum green certificate price for offshore wind	Offshore wind FIT: €10.7 cents/kWh for 10 years; €9 cents/kWh for further 10 years	
Finland	National incentive payment	Wind: €0.69 cents/kWh (+ wholesale tariff) Others: €0.42 cents/kWh (+ wholesale tariff)	Capital grants up to 40% of capital costs
Spain	Incentive payment (50–100MW) National FIT (>50MW)	Wind incentive: €2.93 cents/kWh (+ wholesale tariff) Wind FIT: €7.32 cents/kWh for 20 years, €6.12 cents/kWh after	
Sweden	Federal green certificates	All: ~€1.5–4.2 cents/kWh (+ wholesale tariff) for 15 years	State FITs
US	Production tax credit	Wind: €1.5 cents/kWh (+ wholesale tariff) Marine: €0.8 cents/kWh (+ wholesale tariff)	State-level renewable portfolio standards

Table 4.1 continued

Country	Key Project Support Measures	Price-based Incentive (€ c/kWh)	Other Incentives
China	National renewable obligation State-level incentive schemes		
Taiwan	National FIT	Not yet defined	
Portugal	National FIT	Offshore wind FIT: €7.4 cents/kWh for 15 years Wave FIT: €7.6–26 cents/kWh	Subsidies payments Tax incentives

Note: England and Wales only; more significant financial support is in place for Scotland.

Establishment of support policies specific to ocean energy is being driven by strong targets for overall renewable energy deployment across many countries. In the European Union (EU), for example, the EU Renewable Energy Directive sets an overall target of 20 per cent renewables (across electricity, heat and transport) by 2020. EU member states have different targets according to their current installed capacity, resource availability and gross domestic product (GDP). They are responsible for how their national targets are met across the different sectors.[1] This is a strong driver for increasing renewable energy policy support in EU member states. In 2009, the EU's European Economic Recovery Plan dedicated €255 million of its total €565 million to offshore wind, across five offshore wind farms (see Table 4.2).[2]

International research into offshore wind energy technical and policy issues includes work under task 23 of the International Energy Agency's (IEA's) Implementing Agreement on Wind Energy Systems. Such work is split into sub-task 1 with Risø National Laboratory in Denmark that is focusing on experience with critical deployment issues (ecological issues and regulation; electric system integration; and external conditions, layouts and design) and sub-task 2 with the US National Renewable Energy Laboratory (NREL) that is focusing on research for deeper water. There are also collaborative research and development (R&D) projects internationally, such as the Wind Energy Technology Platform (TPWind) that builds collaboration among Europe's wind industry and public-sector participants.[3]

Table 4.2 EU Economic Recovery Plan

European Economic Recovery Plan – offshore wind farms	€ million
BARD 1: Production of innovative triple foundation system and production and installation of innovative cable in-feed system for a 400MW offshore wind farm.	53.10
Global Tech I: Gravity foundations – gravity foundations for deep-water wind farms using efficient serial manufacturing and fast installation processes.	58.55
Nordsee Ost: Installation of 6MW wind turbine generators (jacket foundation structures) in challenging offshore circumstances, including innovative logistics and installation concept.	50.00
Borkum West II: Installation of innovative 5MW wind turbine generators on tripod foundations.	42.71
Aberdeen Offshore Wind Farm: wind deployment centre – development of a facility for testing of multi-megawatt turbines with innovative structures and substructures and optimization of manufacturing capacities of offshore wind energy production equipment.	40.00
Thornton Bank: optimized logistics for up-scaling the far-shore deep-water Thornton Bank wind farm and demonstration of innovative substructures (jacket foundations) for deep-water offshore parks.	10.00

Source: EWEA (2010a)

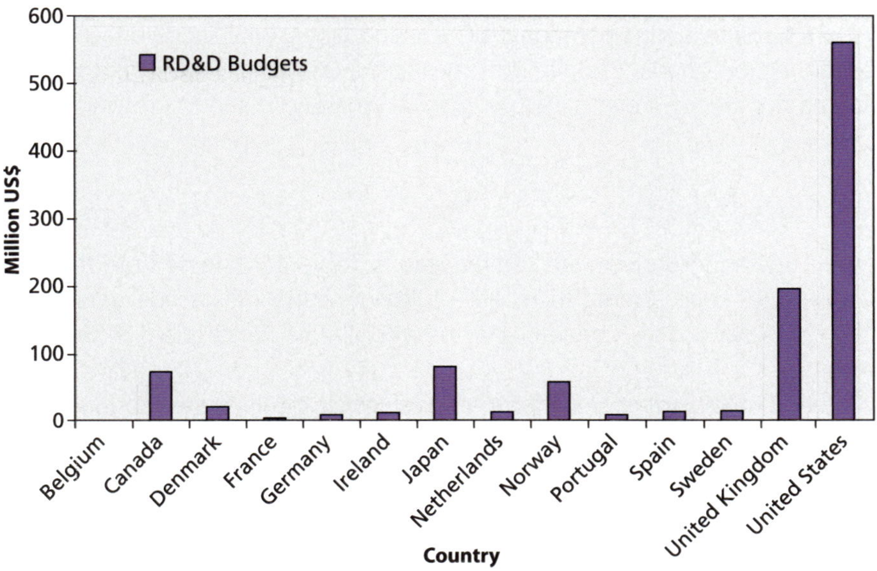

Figure 4.1 Reported government ocean energy RD&D budgets in IEA member countries, 1974–2009

For ocean energy, research, development and deployment (RD&D) also appears to be a very strong component of government policies, particularly due to the earlier stage of development of the industry. Even as early as the period of 1974 to 2003, IEA governments allocated about US$0.8 billion (at 2004 prices and exchange rates, approximately €570 million at current exchange rates) to RD&D into ocean energy technologies. The split across the different governments is shown in Figure 4.1, showing the UK and US as early leaders in terms of RD&D investment.[4] There is growing evidence for on-going and increasing RD&D spending in clean technologies, in general, and offshore renewable technologies, in particular.

4.2 Deployment to date

The present offshore wind industry is located almost entirely in Northern Europe, in the North, Baltic and Irish Seas, surrounding countries where there are less onshore wind resources than in places such as North America and China. Furthermore, deployment is being focused on areas of the continental shelf (shallower seas) as water depth is a principal cost factor in offshore development.[5] The world's first plant, in shallow water, was installed in 1991, about 3km off the Danish coast. Currently, 17 wind farms are under construction in European waters, totalling more than 3500MW. In addition, a further 52 offshore wind farms in European waters have been fully approved, totalling more than 16,000MW. The expected amount of offshore wind energy installed in 2010 was exceeded and amounted to 1444MW; as a result, the cumulative wind capacity installed offshore currently amounts to 3554MW.[6] As well as Europe, offshore turbines are in operation off China and Japan, while additional projects are planned in Canada, Estonia, Norway and the US59. For marine renewables, a limited number of projects exist worldwide typically limited to prototype/pilot projects.

4.2.1 Current projects

Current operating projects are summarized in Table 4.3 (please note that the table excludes tidal barrages such as the 20MW Annapolis Royal tidal generating station in Nova Scotia, Canada, and the 240MW La Rance plant in Brittany, France).

Details of deployment of offshore renewables for the illustrated countries can be found in Appendix E ('Country profiles').

Table 4.3 Operating ocean energy projects by country (2009)

	Offshore Wind		Marine Energy	
Country	Number of Projects	Installed Capacity (MW)	Number of Projects	Installed Capacity (kW)
Canada	–	–	2	1065
Denmark	9	639.15	1	5.5
France	–	–	–	–
Germany	4	42	–	–
Ireland	1	25.2	–	–
Italy	–	–	1	20
Japan	1	1.2	–	–
The Netherlands	4	246.8	–	–
Norway	1	2.3	2	850
UK	12	882.8	4	2250
Belgium	1	30	–	–
Finland	1	24	–	–
Spain	–	–	1	40
Sweden	5	163.65	–	–
US	–	–	–	–
China	1	9	–	–
Taiwan	–	–	–	–
Portugal	–	–	1	400

Notes

1 OJEU (2009).
2 EWEA (2010a).
3 IEA Wind (2010).
4 IEA (2006).
5 IEA (2009a).
6 BTM Consult Aps (2011).

5

Economics of Offshore Energy Projects

This chapter introduces the concept of technology maturity and technology readiness levels (TRLs) and goes on to compare the capital expenditure (CAPEX), operational expenditure (OPEX) and cost of energy of the different technologies, including the cost structures and drivers and variations by country.

5.1 Level of maturity of offshore technologies

5.1.1 Introduction to technology readiness levels

The level of maturity of a technology is one of the key factors to be considered by project developers and financial institutions as part of their investment process, and by policy-makers and government as part of national energy policies.

Methodologies have been developed in order to facilitate an objective assessment of the level of maturity of evolving technologies against a standard set of objective and transposable criteria or achievements. The technology readiness levels (TRLs) methodology, developed by the US National Aeronautics and Space Administration (NASA) for its space programme,[1] has been widely used across many industry sectors – in particular, defence[2] and, more recently, to evaluate low carbon technologies either in their original or in an adapted format.[3]

As described by Mankins (1995): 'Technology Readiness Levels (TRLs) are a systematic metric/measurement system that supports assessments of the maturity of a particular technology and the consistent comparison of maturity between different types of technology.' In its original format (Figure 5.1), the TRLs methodology uses a scale based on nine levels, from 'basic principles observed and reported', corresponding to level 1, to 'actual system *flight proven* through successful mission operations', corresponding to level 9. As indicated above, the definition and number of levels can be adapted to a specific sector, such as wave energy.[4]

It must be noted that the TRLs scale focuses on the technological maturity of the innovation in question. To evaluate a technology's general potential, such a

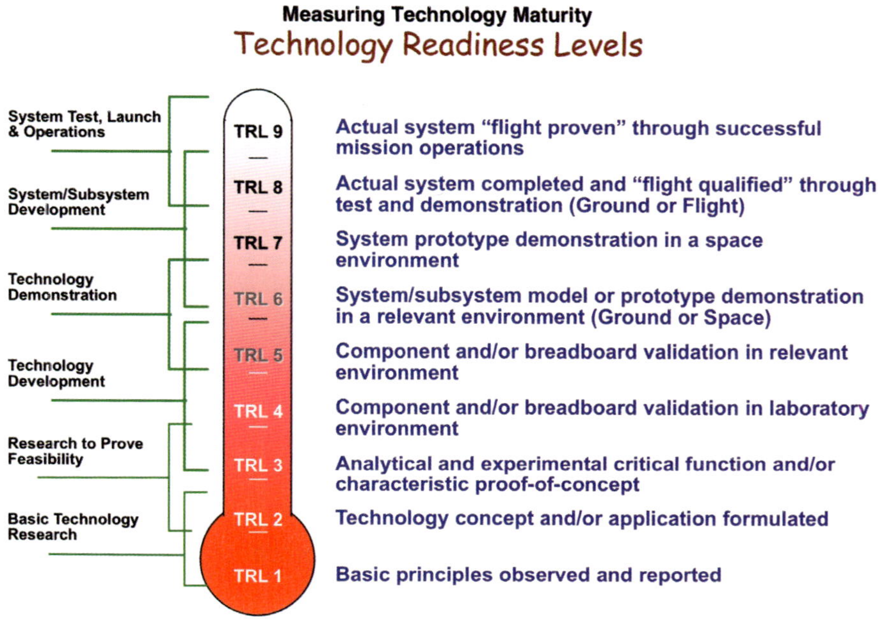

Figure 5.1 Technology readiness levels
Source: Mankins (1995), Technology Readiness Levels: A White Paper

technical assessment should be supplemented with market considerations. Having a particular technology ready for the market does not mean that the market is ready for the technology, as in many cases manufacturing capabilities and general supply chain support, market acceptance, adapted regulation, financial incentives and other conditions will have to be fulfilled before a market breakthrough can possibly occur.

5.1.2 TRLs of offshore renewable energy technologies compared to some fully mature technologies

Table 5.1 shows an assessment of the relative level of maturities achieved by offshore renewable energy technologies (offshore wind, wave, tidal), compared to fully mature technologies reference cases (onshore wind, gas turbines).

Onshore wind and gas turbines are technologies which are both mature, commercially successful and have achieved a high level of manufacturing maturity (lean production processes, stable designs).

In comparison, offshore wind can be considered, as a whole, as a technology which in some cases has achieved or is approaching TRL 9, particularly when conventional wind turbine generator (WTG) and foundation designs have been used for deployment in shallow water depths

Table 5.1 Comparative table of achieved TRLs by various technologies

Technology	Technology Readiness Level (TRL)	Cumulative Worldwide Installed Capacity in MW (2009)
Offshore wind[1]	Up to 9	2110MW
Wave energy	Up to 7	Less than 2MW
Tidal energy[2]	Up to 7	Less than 3MW
Onshore wind	9	160,084MW
Gas turbines[3]	9	1,168,000MW

Notes:
1 Offshore and onshore wind (BTM Consult Aps, 2010).
2 Only in-stream tidal technologies are considered here, and tidal barrage technologies such as La Rance (240MW) or Annapolis (20MW) are excluded.
3 ETSAP (2010).

using standard installation vessels. However, in a number of other cases, lower TRLs should be assigned as:

- many offshore wind turbine models have been produced in very small numbers and have only a few years, at most, of operational experience (and certainly no operational experience over the full, typically 20 years, design lifetime); or
- if a particular sub-system is yet unproven in the field (e.g. using new foundation designs or new types of installation vessels and procedures); or
- if the deployment is to take place in more challenging offshore conditions (i.e. in large water depths (40m or more), at large distances from shore).

When even more radical designs or technologies are considered, such as floating or vertical axis offshore wind turbines, lower levels of TRL (7 for demonstration project, or less) should be assigned.

The most advanced wave and tidal technologies can be judged to have achieved TRL 7, corresponding to the deployment in real seas of full-scale demonstration projects generally consisting of single prototype devices. There is currently no grid-connected, operational wave or tidal farm in commercial exploitation worldwide;[5] but there are a handful of grid-connected single devices[6] at full or large scale and many more technologies in development at various scales across the world.[7,8]

As shown in Table 5.1, there is a clear correlation between increased TRLs and cumulative worldwide installed capacity. This is not surprising, as technologies that have not reached the highest level of maturity will tend to be considered as high risk by investors and project developers. These technologies are also likely to be associated with high capital costs as a consequence of low production rates, characterized by an under-developed supply chain and manufacturing techniques which have not achieved lean production practices.[9] As a consequence, such technologies are likely to require significant market pull through policy and financial support in order to emerge in the power generation sector and compete with established mature technologies.

5.2 Comparison of CAPEX, OPEX and cost of energy

Table 5.2 presents the typical ranges of capital expenditure (CAPEX), operational expenditure (OPEX) and cost of energy values[10] compiled from a number of different sources.[11–15]

Table 5.2 Typically observed or estimated CAPEX, OPEX and cost of energy values by technology

Technology	CAPEX (€/kW)	OPEX (€/kW/year)	Levelized Cost of Energy (€/MWh)	Capacity Factor (%)
Offshore wind	2800–4400	60–100	120–250	35–45
Wave energy	5200–10,800	n/a*	140–530**	25–35
Tidal energy	5800–9600	n/a*	110–220**	35–50
Onshore wind	1200–1800	22–42	60–120	20–35
Combined cycle gas turbine (CCGT)***	620–850	19–26	85–110	50–90

Notes:
* No publicly available figures based on operational data have been identified.
** The cost of energy (COE) estimates provided for wave and tidal technologies are meant to be representative of future first small farms and do not represent current COE associated with one-off prototype demonstration projects (see section 5.5 for details).
*** Mott MacDonald data.

For each technology listed in Table 5.2, typical values have been given as a range rather than a median value. This choice was made in order to highlight the large variations sometimes observed as a consequence of important differences within each technology stream caused by factors such as technical design, project location and others, as discussed in subsequent sections.

Levelized costs of energy, which are themselves a function[16] of capital costs, operational costs, financial costs and site-specific energy production figures (for intermittent renewable energy technologies), can consequently display a large range of values. For conventional thermal plant, fuel costs also need to be included, with the corresponding uncertainty associated with the evolution of such costs in the future.

In the following sections, typical CAPEX, OPEX and cost of energy cost structures of offshore renewable energy technologies are presented and their associated drivers briefly discussed.

In section 9.15.1 (Chapter 9), the levelized costs are compared with the support mechanism financial levels found in different countries.

5.3 CAPEX cost structure and drivers

The CAPEX cost structure of offshore wind projects can be assigned to the following main categories:

- generator (offshore wind turbine, wave or tidal energy converter) supply and installation costs;
- foundation supply and installation costs;
- electrical infrastructure supply and installation costs (cables, sub-station, grid connection); and
- project development and management costs (including permitting, advisory services and studies, land or lease option or purchase).

Table 5.3 highlights the typical CAPEX breakdown currently observed for offshore wind energy projects.

Table 5.3 Typical offshore wind project CAPEX breakdown

Category	Share of Total Cost
Wind turbine generator (WTG) supply and install	50%
Foundation supply and install	25%
Electrical supply and install	15%
Project development, management and miscellaneous	10%

Source: Renewables UK, 2009

The categories listed in Table 5.3 are directly linked to the contractual structure of offshore energy project, where specific contracts will be placed with specialist suppliers. Each of these main categories can be further broken down into a number of sub-categories, such as vessels, raw material, etc. Some of these sub-categories contribute to more than one of the main categories – for example, steel will affect the cost of both WTG and foundations, whereas copper will mainly contribute to the cost of the electrical supply contract and, to a lesser extent, to the WTG costs.

There are currently few data and studies available presenting similar information for wave and tidal projects.[17,18] Similar cost structures to offshore wind projects can, in general, be expected, although the exact breakdown is likely to differ.

The key risk drivers and their relative impact upon the economics of offshore technologies are further discussed in Chapter 6 of this book.

Case study Historical evolution of offshore wind farms' CAPEX

Figure 5.2 presents the observed evolution of offshore wind farm CAPEX.

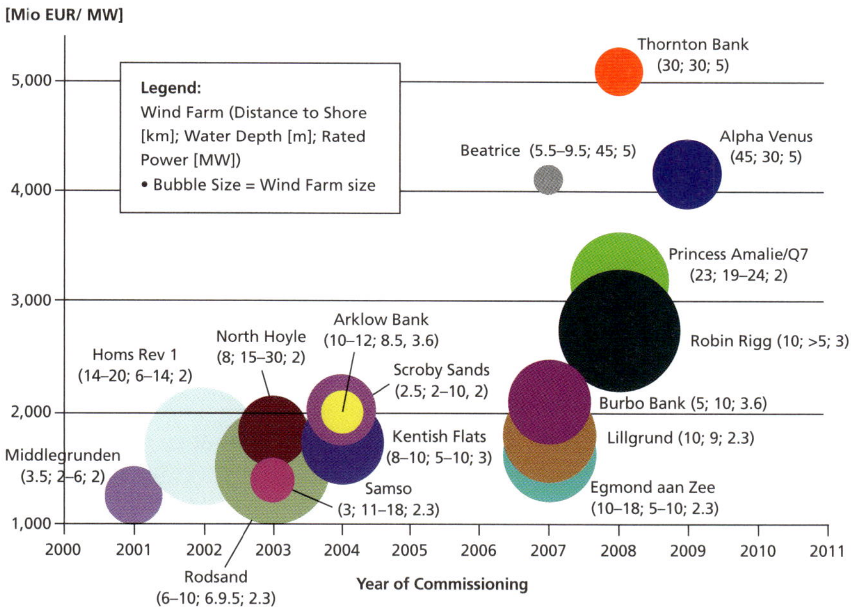

Figure 5.2 Historical evolution of offshore wind farms CAPEX

A number of factors have contributed to the observed sharp increase in the capital costs of offshore wind energy projects between 2000 to 2005 and 2010, including (amongst others, in no specific order or importance) development in more challenging sites (further offshore and in greater water depth); reduced competitive pressures creating a 'seller' market (small number of manufacturers and limited production capacity); onshore wind boom; supply chain pressures from within and outside the sector (shortage of suitable installation vessels, competition from the oil and gas sector); increased raw material costs (steel, commodity); changes in contract structure from single to multi-contracts; high-profile technology failures; and consequent risk re-evaluation by suppliers.

The decreasing capital costs expected to occur with increasing installed capacity in any industry (industry learning curve) have not yet been witnessed within the offshore wind energy industry compared to what happened historically for the onshore wind industry. This is, however, likely to take place in the forthcoming years as a consequence of both 'evolutionary' costs improvements (e.g. from the development of a dedicated offshore wind supply chain) and potentially some 'revolutionary' cost improvements as new, more efficient technologies and designs are introduced.

5.4 OPEX cost structure and drivers

The OPEX cost structure of offshore wind projects can be assigned to the following main categories:

- operation and maintenance costs (scheduled and unscheduled maintenance, including repairs);
- leases;
- insurance premiums;
- provisions (for decommissioning, strategic maintenance reserve accounts);
- site administration and management charges; and
- grid usage charges (country dependent).

Based on the operational experience from a sample of UK wind farms,[19] OPEX costs are shown to be dominated by operation and maintenance costs (€72/kW/year), which represent 76 per cent of the total annual OPEX (€95/kW/year). As most of the offshore wind capacity has been installed in the last few years, the majority of offshore wind farms (and, in particular, the larger, more recent projects) are still under warranty and their operation and maintenance are performed by the wind turbine manufacturers. It is envisaged that even after the expiry of the initial warranty period, the wind turbine manufacturers will continue to provide O&M services; however, this situation may change in the next five to ten years as third-party service companies are likely to be set up to support the industry.

OPEX costs for wave or tidal projects are very uncertain as a consequence of the lack of operational experience, and figures can only be estimated based on best assumptions from the technology developers. As for CAPEX costs, large differences are likely to emerge according to the type of technology and its deployment zone (shore based, near shore or offshore), although keeping O&M costs down is at the forefront for all wave and tidal energy technology developers.

5.5 Cost of energy

Typically observed (for combined cycle gas turbines (CCGT), onshore wind and offshore wind) or expected (for wave or tidal), levelized costs of energy for offshore renewable energy technologies have been provided in Table 5.2.

Wave and tidal costs of energy need to be considered as very indicative as they are more likely, at present, to represent estimates for near future pre-commercial projects, based on best assumptions, rather than the current cost of energy associated with deployed demonstration projects. This is likely to explain why the costs of energy quoted in Table 5.2 for wave and tidal energy appear surprisingly close to those of offshore wind, despite the difference in achieved level of maturity and installed capacity.

Cost of energy for offshore wind energy is also somewhat uncertain and can vary. Although final CAPEX costs are known once the project has been procured, installed and commissioned, uncertainties in OPEX costs only reduce as experience is gained, while capacity factors will be heavily dependent upon the actual availability achieved by offshore wind farms. For example, best estimates for the cost of energy of UK offshore wind farms have been revised from €110/MWh in 2006 to €173/MWh in 2009 as a consequence of increased CAPEX and OPEX costs.

5.6 Differences between countries of project location

5.6.1 Conditions in different countries

The geographical locations of offshore renewable energy projects influence their economics/attractiveness of investment. These factors can be broadly grouped into two main categories:

1. Factors associated with local project conditions, such as:
   - resource (wind speed, wave regime, or tidal flow velocity);
   - bathymetry (steepness of increasing water depth);
   - seabed conditions (geophysical nature of seabed and impact upon foundation or mooring options); and
   - proximity to infrastructure such as grid connection or ports.
2. Factors associated with countries, such as:
   - planning, environmental, grid connection, legal and consent regime, and requirements;
   - existence of incentive mechanisms and targets; and
   - political will and leadership.

The first set of factors (resource, bathymetry, seabed conditions and proximity to infrastructure) contributes to making the North Sea the main area for offshore wind deployment. Other countries or regions, such as the US West Coast, the West Coast of Ireland and Norway, characterized, for example, by less favourable bathymetric conditions (deeper water close to shore), are consequently considering different technologies such as wave or floating offshore wind turbines.

It is beyond the scope of this book to fully explore and discuss in detail how the complex interactions of the second set of factors influence the overall economics of offshore energy projects; however, the issue is briefly discussed in the section below.

5.6.2 Value of political will and leadership for project costs and risks

Currently, the largest markets for offshore wind deployment are Belgium, the UK and Germany, countries which have shown strong political leadership in order to gradually remove or reduce the barriers[20] to development, while at the same time tailoring the financial incentives (multiple renewable obligation certificates (ROCs) in the UK, tariffs linked to project's water depth and distance to shore in Germany, grid subsidies and tariffs in Belgium) necessary to ensure the economic viability and attractiveness of offshore wind energy projects.

5.7 Conclusions

Utilizing the technology readiness levels (TRL) methodology, it is clear that all marine technologies have some way to go before being considered 'mature'. It is also evident that offshore wind is more developed than wave and tidal, which translates into a higher installed capacity worldwide, as well as more certainty regarding CAPEX, OPEX and cost of energy.

Cost structures for CAPEX and OPEX are expected to be largely similar for all marine technologies; however, the exact breakdown is expected to vary. The actual location of deployment could be a significant factor in terms of the physical characteristics of the site. Regarding the political landscape, countries that have shown strong political leadership and tailored financial incentives, such as Belgium, the UK and Germany, appear to be leading the way in offshore wind.

Notes

1 Mankins (1995).
2 Technology readiness levels (TRLs) have been used for many years by both NASA and the US Department of Defense to develop advanced mission-critical systems.
3 TRLs methodology is, for example, used by the Energy Technology Institute for their technology programme (see http://www.energytechnologies.co.uk/Home/

Technology-Programmes/How-We-Operate/Types-of-Projects.aspx, accessed 8 July 2010), or by the US Department of Energy for their marine and hydrokinetic technology readiness advancement initiative (see https://www.fedconnect.net/FedConnect/?doc=DE-FOA-0000293&agency=DOE, accessed 8 July 2010).

4 Waveplan (2009).

5 The first commercial wave farm, the 2.25MW Aguçadoura project, consisting of 3 Pelamis P-750kW wave energy converters and commissioned at the end of 2008 in Portugal, is not currently operational following the financial collapse of owner Babcock & Brown and failure to identify a new owner to date.

6 Limpet 500kW oscillating water column wave energy converter, Islay, UK; MCT's Seagen 1.2MW tidal energy converter, Strangford Loch, Northern Ireland, UK. A number of wave and tidal devices are also undergoing temporary deployment at the grid-connected wave and tidal test sites of the European Marine Energy Centre, Orkney, UK.

7 Waveplan (2009).

8 Deutsche Windguard (2010).

9 See, for example 'manufacturing readiness level' terminology, serving a similar purpose for manufacturing as the TRL terminology for technology (en.wikipedia.org/wiki/Manufacturing_Readiness_Level).

10 When referring to costs provided in other currencies, conversion to Euros has been done using the following rates (based on 13 July 2010 rates listed on www.xe.com) rounded to 2 decimals: US$1 = €0.79 and UK£1 = €1.20.

11 DECC (2010c).

12 Dudziak (2009).

13 Milborrow (2010).

14 DECC (2009).

15 Carbon Trust (2010).

16 DECC (2010c).

17 Carbon Trust (2006).

18 FREDS Marine Energy Group (2009).

19 DECC (2009).

20 A description and analysis of technical and non-technical barriers is presented in Chapters 8 and 9.

6

Project Risks and Related Project Costs

6.1 Introduction

There are a number of risks related to offshore projects. Complete removal of risks is uncommon. Mitigation measures are, however, likely to reduce these risks to acceptable levels in terms of project economics. Depending upon the nature of the risks, these mitigation measures can, for example, be a consequence of policy interventions from government or of a technical or financial nature.

6.2 Effects of project risk assessment on economics

6.2.1 Perceived risks and evident risks

'Evident risks' represent the real well-founded risks that occur in the implementation of offshore renewable energy projects. 'Perceived risks' derive from the assessment of those evident risks by the involved stakeholders and vary according to their perspective.

Offshore renewable energy technologies, similarly to every other power generation alternative, are exposed to design, construction and commissioning risks, independent of their respective level of maturity. The uncertainty associated with construction costs derives from their exposure to:

- commodity prices volatility, which directly affects the cost of raw materials and indirectly the production cost of technical components (i.e. turbines, blades, nacelles); and
- supply chain implications such as equipment manufacturing bottlenecks, which may postpone the completion of a project as well as lead to price escalations. As a consequence, such supply chain issues may postpone or eliminate revenue generation. Therefore, supply chain accounts for both a development risk and a revenue risk for a project. Supply chain risk is generally out of the control of the project developer and does not qualify for short-notice resolution.

Another evident risk accounts for uncertainty due to weather. Bad weather is the most unforeseeable and most obstructive problem in offshore projects. For instance, in the case of a wind farm, offshore wind speed is likely to hinder the installation of towers, nacelles and rotor blades. Moreover, high waves and rough waters have the potential to disrupt any part of the wind farm installation process. This translates into project completion delays and, hence, adds to development risk.

As already implied, the maturity level of a technology can very rarely be insulated from the supply and construction cost components variation. The noticeably variable trend of combined cycle gas turbine (CCGT) prices over the last decade, driven by the introduction of new models, demonstrates that even well-established technologies are not immune from supply and construction price fluctuations.

Although the maturity stage of a technology does not eliminate the supply and construction price risks, the less mature technologies will typically incur additional costs, known as a first-of-a-kind (FOAK) premium, versus the mature or nth-of-a-kind (NOAK) technologies (DECC 2010c, sections 3.6 and 3.7). This FOAK premium will relate to impacts in the construction process, the design, and the manufacture of components. However, it should be noted that frequently, manufacturers have taken the strategic decision to offer FOAK at a discounted price to enable them to enter the market. In cases where such introduction to the market has been successful, subsequent transactions are at a premium, driven by a demand for new technology that was proven by FOAK.

Accounting for early-stage technologies, offshore projects do not yet feature a broadly standardized construction process, which results in an additional risk perception. The industry is still aiming to improve the construction process, an indicative example being the different approaches regarding the degree of onshore and offshore assemblies of components.

In terms of design, issues such as resilience of materials against extreme weather conditions, performance of technical components in an offshore setting, and adequate testing of installable equipment raise project outturn uncertainty.

As more experience is acquired in manufacture, further optimization of the production line of equipment is expected. As a technology matures, cost efficiencies and potential contingencies are alleviated regarding issues such as the input of and processing of materials, distribution of output, staff and supply chain management.

Nevertheless, as indicated above with the FOAK example, commercial issues are likely to have a big impact upon the cost of supply and construction, such as level of competition, barriers for new entrants to entering the market and number of projects (buyers), etc.

6.2.2 Share of project costs due to risks

In principle, there are four methods for dealing with the risks associated with any construction project.

The first method for the developer would be to pass the risk to another party, via a single supply and construction contract with a contractor, who would be willing to undertake the whole completion risk in exchange for a price premium added on the contract price. The choice of different project schemes (single versus multi-contract) does not alter the real overall cost or risk profile of the project; it simply affects the allocation of risk between the developer and the contractor(s). However, currently, single-contract schemes are not feasible for offshore projects as no contractor is prepared to take on this risk for a reasonable price, from the developer's point of view.

The second method would be to insure against the risks; however, there are no insurers who are prepared to underwrite such risks at present.

The third option involves 'designing the risk out'. There are two aspects to designing out the risk: studies, schedules and applying lessons learned from other industries with similar experiences (e.g. offshore oil and gas). Good-quality extensive studies should provide a good appreciation of the site, leading to a more appropriate design for the site. Some conservatism will be applied in any design to account for reasonably unforeseen ground conditions. In terms of schedule, sufficient slack should be incorporated to enable the project to meet the deadlines despite potential delays due to weather-related implications or lack of coordination between different contractors. These precautions taken by the developer should include contingency-oriented scheduling of the overall project completion, such as flexible time slots between complementary component processes. However, there will be a substantial cost premium from such modifications.

The fourth method is to add a contingency item to the budget. This risk premium is intended to keep the developer within the initial estimated overall budget and should be particularly high for relatively 'immature' technologies. However, it does not fully eliminate the possibility of an even higher out-turn project cost. The actually chosen value of the respective contingencies is associated with the perceived risk to the financiers. Experience indicates that project financed schemes have a typical contingency provision of up to 10 to 15 per cent (depending upon the risk profile of the project, as well as the number of supply and construction contracts). Utilities are likely to have lower requirements; but it depends upon the risk perception within each company.

From a borrower's perspective, uncertainty concerns over a project can alternatively be reflected in applied loan interest rates. However, 'double-counting' of risks should be avoided. In other words, caution should be taken so that contingencies and risk-adjusted loan interest rates do not incorporate the same perceived risks.

In practice, combinations of the above stated methods are often used when designing and planning offshore renewable energy projects.

6.3 Key technical project risks

6.3.1 Overview

Key technical project risks can be differentiated into two main categories:

1 completion risks, defined as the risk of a project not achieving successful commercial operation; and
2 operational risks, defined as the risk of a project not being able to operate successfully and to generate revenues sufficient to meet its business needs.

Tables 6.1 to 6.6 briefly summarize typical risks among these two key categories, the owner of the risk and effective mitigation measures. It should be noted that depending upon the particularities of each project, some of these risks may not be relevant and, equally, further risks may be introduced. The risks have been presented and discussed for offshore wind projects; but the vast majority of these will also apply to wave or tidal energy projects. Key differences between wave and tidal energy projects and offshore wind projects are discussed in section 6.3.4.

6.3.2 Completion risks

Table 6.1 Risks associated with ground and environmental conditions and permitting

Risk	Risk Owner	Mitigation
Extent of site investigations	Project owner	A comprehensive understanding of the site conditions for a project is key to mitigating completion risks. Investigations should include sea bed conditions, and weather and marine environment, as well as delivery of equipment and the capacity of the project port base.
Achievement of all necessary permits	Primarily project owner	Early engagement with stakeholders (including the public) and diligence in the timely achievement of project permits is an obvious mitigant. Some permits may need to be obtained by the contractors. If the project is delayed due to a contractor permit, then the risk will probably be covered by damages. Legal/ planning advice typically needs to be sought to confirm the extent of permits required.

Table 6.2 Risks associated with participant capabilities

Risk	Risk Owner	Mitigation
Capability of sponsors	Project owner	Without strong sponsor capability and adequate leadership, progress is likely to be slow, inconsistent and expensive. The only mitigants are to bring in new investors or to hire capable advisers.
Capability of project managers	Project owner	An offshore wind farm developed and constructed under current models is a sophisticated process with complex interfaces needing strong management. Furthermore, dealing with a multi-contract structure as typically used for offshore wind farms needs an experienced and capable construction management. The mitigant is a capable team.
Capability of contractors	Project owner/ contractor	There are a number of capable contractors covering the various tasks needed to deliver an offshore wind farm; but in addition to corporate experience, the experience of individuals charged with delivering the project is also a key mitigant. A contractor's performance needs to be backed by a strong contract and damages for poor performance.

Table 6.3 Risks associated with design

Risk	Risk Owner	Mitigation
Responsibility for information and design	Project owner/ contractor	Depending upon the contract, the responsibility for information provided may rest with the contractors or the owner. Diligence in project planning is required to mitigate this risk as far as possible. Design responsibility typically lies with the contractors. Third-party verification is a key mitigant.
Turbine technology risk	Contractor	A technology risk has two aspects: (i) will it work; and (ii) will it continue to work? While operational experience is the preferred mitigant, owners often want to benefit from the latest development that may be bigger, better and more efficient. Independent analysis and certification of a design can give some comfort; but the key mitigant must be a manufacturer's warranty with liquidated damages for poor performance.

Risk	Risk Owner	Mitigation
Selection of foundation type	Contractor/ project owner	With larger units now being the norm for offshore wind farms, the selection of foundation type may be affected not just by sea bed geotechnical conditions, but also by the availability of plant and equipment to install the foundations and the experience of contractors working in the required conditions. The mitigants are diligence during the planning phase, contingency planning and damages payable by the contractor.
Grid code or interconnection agreement compliance	Contractor	Grid code compliance is a risk that must be held by the supply and construction contractors. Depending upon the contract structure, grid code compliance may need to be guaranteed by more than one of the contractors. In the case of a split responsibility, independent analysis needs to be performed to ensure that each component will satisfy code compliance.
Coordination of design with O&M requirements	Project owner	Lack of coordination between the design and O&M team of the owner is not uncommon. One area requiring particular effort for an offshore wind farm covers the design basis and detail design of the boat landing for the turbine towers. This coordination needs to consider the approach to be followed for O&M and selection of the maintenance boat (or other means). It also needs to consider the wind farm target availability and expected wind and sea conditions.

Table 6.4 Risks associated with performance

Risk	Risk Owner	Mitigation
Offshore wind park layout	Project owner	The wind park layout needs to be such as to achieve the design life of the turbine and other key components. Issues can arise due to turbulence and fatigue loading if, for example, the turbine spacing is too close. The wind turbine generator (WTG) supplier should be consulted early to confirm that the layout is appropriate for the proposed WTG model. There are also potentially grave planning implications.

Table 6.4 continued

Risk	Risk Owner	Mitigation
Selection of appropriate turbine classification	Contractor	Care needs to be applied to selecting the correct class of turbine for the expected wind and meteorological conditions. The impact of poor selection could be a reduced life of blades and other components or possibly a reduction in energy yield.
Turbine performance	Contractor	Care needs to be taken in drafting the turbine supply and O&M contracts to ensure that the necessary information will be available to carry out turbine performance tests. The test code typically referenced (IEC) needs careful interpretation and information from an independent met mast must be available to carry out a successful power curve test. However, the IEC requirements are almost impossible to implement. Therefore, provisions for alternative practical methods (independent party assessment via a pre-agreed methodology) should be agreed between the parties. Liquidated damages should be available for poor performance. It should be noted that the turbine performance is likely to be covered under the O&M agreement because of the time typically needed to carry out a full performance test.
Achievement of appropriate certification	Project owner/ contractor	Although certification is not a guarantee that a WTG will be free from design defects, it does give some comfort that the design process has gone through an independent review process. Careful attention should be given to confirm that the certification is consistent with the requirements of the project.

Table 6.5 Risks associated with manufacturing and construction

Risk	Risk Owner	Mitigation
Health and safety	All parties	While local legislation may be satisfied, particular efforts should be applied by all project stakeholders in planning the project health and safety in particular covering policies and detailed plans, and covering all phases of the project from construction to operation. It should be clear who is responsible (i.e. organization and safety officer) for safety in each project operation. The safety procedures and actions should be subject to regular audits and feedback.

Risk	Risk Owner	Mitigation
Environmental damage	All parties	There are a number of areas where environmental damage can occur (e.g. noise from piling activities) during both the offshore and onshore activities. The environmental planning should have studied the potential risks and considered suitable mitigation resulting in detailed environmental monitoring and management plans to be drafted and implemented by contractors.
Manufacturing quality	Contractor	There have been some high-profile failures of wind farms due to manufacturing quality. These problems have occurred largely because of a lack of focus on quality assurance by the manufacturer, often in relation to sub-vendors, and have been as a result of over-trading or time pressure during manufacturing. The contract conditions should mitigate the risk under the warranty and damages provisions; but in the long term, the owner is likely to be affected by poor quality unless the contract terms are strong and include substantial damages. To some extent, the risks can be reduced by the owner's project management and independent auditing, especially aimed at areas that have been shown as problematical.
Weather risk	Shared risk depending upon conditions of contract	It is typical for offshore projects that the weather risk will be shared between owner and contractors. The contracts should state the basis regarding weather. A number of options are available – a typical option is for contractor to take 'average' weather. This needs to be defined in the contract so that the number of 'weather days' can be recorded. The contractor gets relief from schedule but no additional costs for weather days.
Cable terminations	Contractor	Particular effort is needed to successfully complete the cable termination within the turbine towers or the substation. Contractors often try to plan the cable-pulling activities without diver intervention. While the risk will be carried by the contractor and mitigated by damages, the owner should ensure that arrangements are in place for divers to assist if required.

Table 6.5 continued

Risk	Risk Owner	Mitigation
Availability or loss of equipment	Contractor	With the use of large turbine units, it must be appreciated that there may be very few items of key construction equipment (i.e. barges, lifting equipment and piling hammers are examples) capable of carrying out a required task. Even if there are alternative items of equipment, these may be committed to other projects for a considerable time. If a key piece of construction equipment is lost just prior to or during project implementation, the only option open to an owner may be to delay the project. Although the risk of construction equipment availability may rest with the contractor, legal advice will be required to understand if the contractor can gain relief because of *force majeure* or other contract terms. A provision for contingency allowance should also be considered.
Contract structure and interfaces	Mostly project owner	In the current market, no contractor is willing to accept the whole supply and construction risk (at a reasonable premium) and projects are being structured with two or more contracts. Great care is needed in confirming that the scope of work is complete and in understanding the interfaces. An interface agreement is an important mitigant, although not all contractors are willing to sign a document also signed by a third party. Schedule float is a further mitigant; but the owner may wish to manage the risk, in which case the schedule should include float to take account of a delay by one party in achieving an interface. In such a case, the mitigant is to provide a contingency sum within the capital budget to cover the cost of an interface delay. It is also prudent to include a contingency to cover any possible shortfalls in scope. Strong project management is also required.
Cost over-runs	Project owner	Cost over-runs can occur in any project; but with more contractors participating in the supply and construction and with more contractual interfaces, the risk increases. The mitigants are diligence in the contract drafting and negotiations, strong project management (in particular, covering the interfaces) and setting a realistic contingency budget.

Risk	Risk Owner	Mitigation
Damage to property	Project owner	A collision between a ship and, for example, a turbine foundation/tower could cause a serious delay. While efforts in managing the offshore activities may help, a mitigant is likely to be insurance. Damage to the interconnection could be due to a number of reasons, either by a wind farm contractor or by a third party such as a vessel anchoring in a forbidden area. Again, a practicable mitigant may be insurance. It is likely that the interconnection cable will cross other sub-sea pipes and cables. Details of these existing services should be collected during a thorough series of investigations and negotiations with permitting authorities and other service owners. The construction contractor responsible for dredging and cable installation should be obliged to carry out a final physical survey prior to commencement of dredging to confirm or otherwise the information collected during the project planning. If possible, the dredging contractor should take responsibility for any damages to existing services.

6.3.3 Operational risks

Table 6.6 Risks associated with project operation

Risk	Risk Owner	Mitigation
Major intervention risk	Contractor	Major intervention (MI) risk is typically preferable to rest with the O&M contractor, at least for a significant part of the O&M term; however, some developers may choose to carry this risk depending upon their strengths. The O&M contractor should have access to barges and cranes at short/reasonable notice, possibly via a third-party agreement. For the contractor, the risk is likely to be the balance between achieving the guaranteed availability against the cost of carrying out an MI. It is likely that the contractor will wait until there are a number of turbines requiring repair. It is important that the liquidated damages for non-availability will reimburse the expected revenues.

Table 6.6 continued

Risk	Risk Owner	Mitigation
Type fault	Contractor	There have been a number of well-reported issues with offshore turbines suffering from type faults. Operating experience is the key mitigant, although in a fast developing market, owners are often keen to benefit from the latest technology, with the result that equipment is often installed with little operating experience. While diligence during the planning phase of the project should help to understand the risks, the key mitigant must be the guarantee, damages and liability limits accepted by the wind turbine generator (WTG) supplier
Availability	Contractor	For a wind farm, high availability is vital in order to maximize revenue. A number of issues affect the availability – the O&M method is likely to have the greatest impact. Care is needed in project planning to ensure that there is consistency between the wind farm design for accessibility and the approach to be followed for visits to turbines. There is a commercial balance between O&M costs and availability; these need to be considered carefully in contract negotiations. Liquidated damages are the key mitigant against poor availability. Finally, care is needed during the contract negotiation of the O&M contract with respect to the definition of 'availability'; turbine suppliers will endeavour to include events that the owner would expect to be considered as non-availability.
Availability of spares	Project owner	The offshore wind industry is new and developing, and there is no guarantee that wind turbine manufacturers will still be around in ten years. Also, with consolidation, some models of wind turbines are likely to be withdrawn. As a mitigant to the risk of spares not being available, effort needs to be put in place to ensure that an owner will have free access to the design of all components so that, if necessary, they can be manufactured by a third party.

Risk	Risk Owner	Mitigation
Energy yield	Project owner	Energy yield is a significant subject in its own right and risks are mitigated by diligence in the energy yield analysis, prudent selection of the yield and consideration of sensitivity cases. The risk could be related to the performance of the turbine (i.e. a power curve issue) in variability in the wind. Inclusion of lower P50s and P90s in a downside scenario is important.
Lack of capable operator	Project owner	Operator capability is important and needs to be addressed by diligence during the project planning process. However, there are few companies with a significant track record, and owners may need to consider capability from other perspectives – in particular, the suitability of operational plans and other plans and the capability of personnel. The risk should be mitigated, where possible, by strong damages for poor performance.

6.3.4 Discussion of risks associated with offshore wind versus wave and tidal

Although the risks highlighted above have been presented and discussed for offshore wind projects, the majority of these will also apply to wave or tidal energy projects. Key differences between wave and tidal energy projects and offshore wind projects are the higher technology risk associated with the former technologies, as well as the lack of a mature supply chain.

Wave and tidal technology risk is demonstrated, for example, by the fact that the lenders are currently unwilling to finance wave or tidal energy projects, and such projects are, for the foreseeable future, likely to remain developed by utilities and balance sheet financed.

6.4 Impact of key variables upon total cost structure

6.4.1 Description of key variables

Figure 6.1 demonstrates key variables, interdependencies and impact upon the overall project cost structure. As shown, all items eventually relate to either project capital expenditure (CAPEX), operational expenditure (OPEX) or revenue structure. It should be noted that the items shown on the left in Figure 6.1 are not necessarily exhaustive, but rather indicate potential issues that are commonly faced during development and implementation of an energy project.

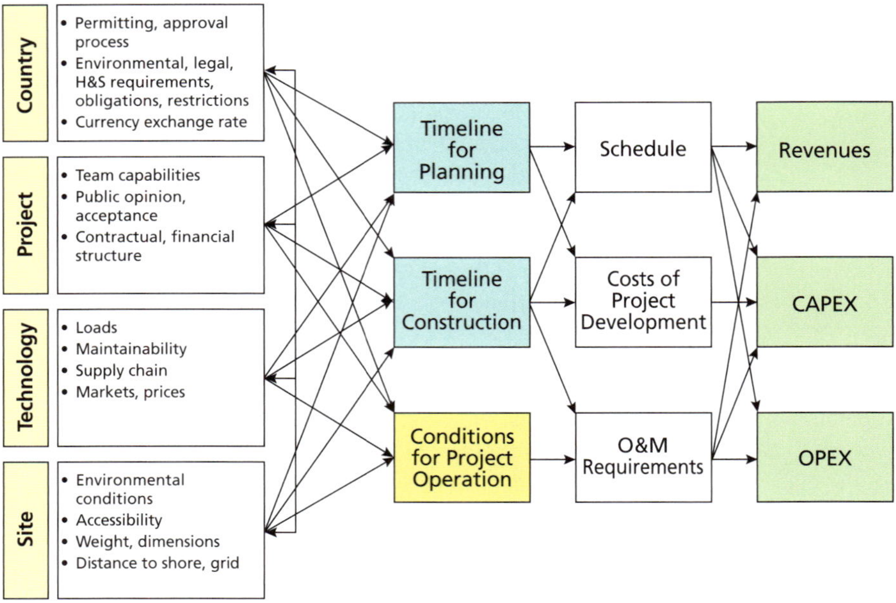

Figure 6.1 Key variables to project cost structure
Source: Mott MacDonald

Issues to do with the country of project location often relate to permitting and approval processes, as well as requirements, obligations and restrictions of environmental, legal or health-and-safety background. Currency and exchange rate are also considered as country-related issues. Project-related issues often comprise capabilities of the project team, public affairs and the proposed contractual and financial structure of the project. Issues associated with technology commonly include supply chain, market conditions and prices, as well as loads and component maintainability, which in turn are often determined by site-specific conditions. Issues related to the respective site comprise environmental conditions, logistical aspects such as constraints regarding accessibility, weight or dimensions of transport, as well as the distance to shore or to the point of interconnection.

While all the issues named above are largely interdependent, they all determine the timeline for project planning and project construction, as well as the conditions for project operation. The timelines for both planning and construction define the project schedule. Both timelines, in turn, have an effect on the cost of project development. Operation and maintenance (O&M) requirements of the project are defined by the conditions for project operation. The project schedule determines the start date of project operation (where the project will start generating revenues). Furthermore, the project schedule affects CAPEX and OPEX. Project CAPEX is also affected by the costs of project development. Likewise, the O&M requirements affect revenues and OPEX.

Delays in implementation of a project, such as a delay of the grid interconnection, often occur due to aggressive timing in a tight schedule for project installation, providing insufficient float within the sequence of works. The cause of delays is therefore frequently related to either project planning or project management.

It is beyond the scope of this book to describe all potential issues and their relevant effects on a project cost structure in detail. Especially with regard to offshore wind, there are studies available in the public domain dedicated to specific issues such as supply chain or analysis of capital cost development.[1] The following case study describes the incentive structure embedded in the German Renewable Energy Act for projects located further offshore and in deeper waters.

Case study Water depth and impact upon overall cost structure and revenues

With increasing distance from a project site to shore, the water depth also usually increases. At offshore wind farm sites around the UK and in the North Sea, there are water depths of roughly 30m to 50m at distances of 30km to 40km offshore. Monopiles as well as jacket foundations have become the most appropriate foundation solution for projects within the above range of water depth from a technical and economic perspective. Along the coasts of Norway or the East Coast of Taiwan, the seabed is a lot steeper and water depths increase already closer to shore, requiring different foundation solutions (e.g. floating structures as currently being tested with the Hywind turbine 10km off the Norwegian coast in about 220m of water depth). The applied mooring system is reported to be suitable for depths of up to 700m.

In regions where the visual impact of an offshore energy project affects its viability, potential projects have to be developed further offshore. The increasing distance has several advantages, higher and more stable wind resource with less turbulence by fewer obstacles being the most significant one. At the same time, increasing distance to shore and increasing water depth have a number of disadvantages. Deeper water requires more challenging foundation solutions, bringing in potential supply chain issues, as well as logistical and installation constraints. The grid connection cable has to be longer compared to projects closer to shore and its installation is likely to be more challenging and more expensive. Both foundation and cable result in higher project CAPEX and potentially higher level of risks. Higher risks associated with project development and installation require a higher contingency budget and, hence, higher overall costs for financing. Transfer of equipment and O&M personnel is more difficult due to exposure to higher waves and is also more time consuming if the project site is further offshore. Maintenance activities are more constrained and interventions with special equipment required are more expensive, increasing overall project OPEX. Furthermore, availability of the installed energy technology is likely to be lower compared to projects closer to shore, which results in lower revenues. All in all, the levelized costs of generation are higher with increasing water depth.

In some countries such as Germany, higher remuneration is granted for projects that are located further offshore in order to incentivize the development of such projects. The revised German Renewable Energy Act (2009) provides a tariff structure with higher remuneration for projects developed further than 12 nautical miles offshore and in more than 20m of water depth. Projects receive a fixed tariff of €13 cents/kWh fed into the grid for an initial period of 12 years. Early action is further incentivised by a bonus of €2cents/kWh on top of the fixed tariff if the project is commissioned before 1 January 2016. The initial period of 12 years is extended by 0.5 months for each full nautical mile that the project is located beyond the 12 mile zone and by 1.7 months for each full metre of water depth beyond 20m. After expiry of the initial period including the extensions, the tariff drops to €3.5 cents as basic remuneration.

6.4.2 Sensitivity analysis

Since an offshore energy project is a complex mixture of components, contracts and technical requirements, etc, it is difficult to quantify the impact that single drivers and mechanisms have upon the overall cost structure of a project, in practice. As a result, a more qualitative view on how such drivers and mechanisms influence project costs is provided. However, as it is possible to observe the effect of single variables in theory, sensitivity analyses have been added to compare various key drivers using a simplified financial model. The underlying assumptions of the base case are provided in Appendix C.

Typical values for the relative impact of price levels as key drivers to project CAPEX are provided in Table 6.7. It is apparent that the currency and related fluctuation of the exchange rate have a major impact upon overall CAPEX and, therefore, upon the overall project cost structure. Installation, linked to labour costs, has the second largest impact of approximately 20 per cent. Fluctuation of material costs for steel and copper has less impact upon the overall CAPEX. An increase of 20 per cent in the steel price does not affect the overall CAPEX by 20 per cent, but rather by an amount multiplied by the specific value content of steel among the total project costs.

As shown in Figure 6.1, project CAPEX is not only influenced by single drivers, but rather by different variables and mechanisms at the same time. CAPEX can be highly influenced by strategic decisions of the WTG or foundation manufacturer in order to introduce a new prototype model in the market. In some cases, project CAPEX may benefit from the main business of the sponsor that might have an entire fleet of installation and service vessels ready built. Therefore, it is difficult to analyse the impact of single drivers and key variables.

Table 6.7 Drivers of project CAPEX

Contract/Category	Total Share	Value Content			
		Installation	**Steel**	**Copper**	**Currency**
Wind turbine generator (WTG) supply and installation	50%	10%	10%	2%	95%
Foundation supply and installation	25%	30%	25%	0%	80%
Electrical supply and installation	15%	40%	5%	10%	70%
Project management	5%	0%	0%	0%	0%
Miscellaneous	5%	0%	0%	0%	0%
Total (value weighted)	100%	19%	12%	3%	78%

Source: Renewable UK (2009)

Project OPEX is influenced in the same way as CAPEX. However, OPEX is more liable to be affected by changes in labour costs and costs of spare parts. Currency remains an issue for OPEX if the entities of the project's company and O&M provider are cross-border.

Sensitivity analyses were carried out to examine the impact of single key variables upon the overall cost structure of offshore renewable energy projects. In the sensitivity cases, the cost of energy from the base case was compared by modifying the variable in question and keeping all other project inputs constant. As explained above, this is a rather theoretical approach as the overall project cost structure is always affected by a mixture of drivers and mechanisms. The approach is intended to indicate the relative impact of each of the variables analysed rather than to provide exact numbers.

As shown in Figure 6.2, the largest impact upon energy costs in offshore wind is due to WTG costs. Changing WTG costs can be related to any of the drivers described above (i.e. costs of material or labour, exchange rate or other reasons). O&M costs also have a large impact upon energy costs; they are mainly driven by labour costs and costs of spare parts. Costs of the installation vessels have a lower impact compared to WTG and O&M costs as they are considered a one-off cost item within the overall CAPEX during project construction.

With regard to wave and tidal energy (see Figure 6.3), a similar impact can be seen in the range and order of the variables affecting levelized costs of energy. As in offshore wind, devices and O&M cost variations have a large impact upon the levelized costs of energy.

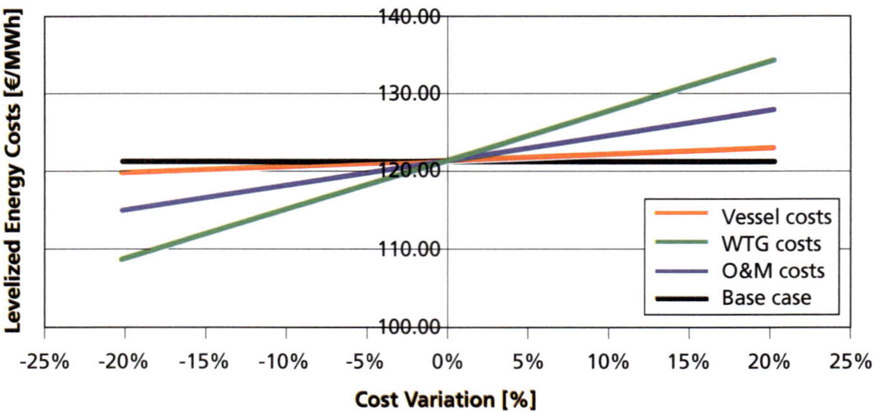

Figure 6.2 Offshore wind cost variation

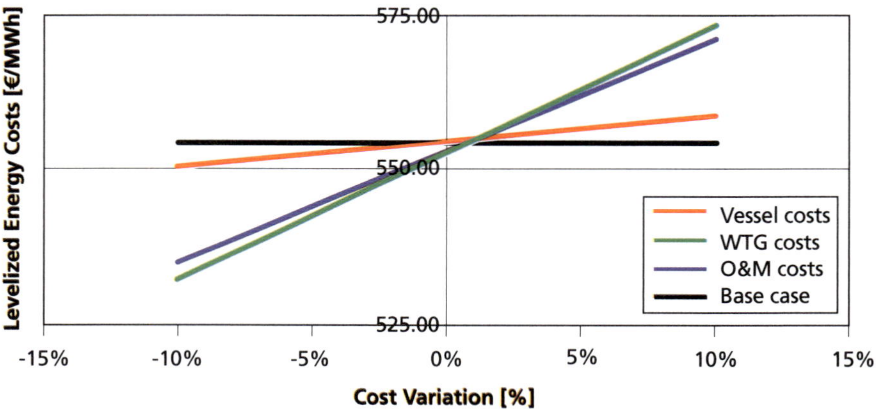

Figure 6.3 Wave/tidal cost variation

6.4.3 Risk management in various countries

There is currently no clear evidence regarding different approaches to risk management from different countries. Instead, each project shows different risk management depending upon the risk that developers are willing to accept and that suppliers are being rewarded for.

6.5 Conclusions

There are a number of risks associated with offshore projects. Complete removal of such risks is unusual; however, mitigation measures can reduce these risks to an acceptable level in order to facilitate project development.

The residual risks have an effect on project economics/investment attractiveness. A number of risks associated with commodity price volatility and supply chain issues have an impact and cannot be fully influenced by the developer. Nevertheless, commercial issues may be more important in influencing the costs of a project. Regarding technical risks, there are mitigation measures that can be used to reduce or even eliminate the impact of completion and operational risks.

In terms of cost structure, there are a number of largely interdependent variables that affect CAPEX, OPEX and revenues. These variables include permitting, environmental issues, exchange rates, contractual structure, supply chain, market and accessibility, to name a few.

Sensitivity analyses were carried out for offshore and wave and tidal projects. For offshore wind, which is a considerably more mature technology, WTG and O&M costs appear to have the greatest impact, whereas for wave and tidal, the effect of the capacity factor, which is still an unknown, has the most significant effect.

Note

1 Renewable UK (2009).

7

Financing of Offshore Renewable Energy Projects

7.1 Financing options

Developers wishing to realize a project have a number of potential financing options, which are broadly categorized as:

- balance sheet financing, where the project is funded by any combination of the developer's own cash reserves and debt finance secured against other parts of the developer's business or assets;
- project financing, typically non-recourse financing, where the project is funded at least in part by debt secured against the future cash flows of the project only, which is established as a standalone entity.

Each option has advantages and disadvantages that may affect their suitability and availability for the financing of offshore renewable projects. These are investigated further in the following sections.

7.2 Balance sheet finance

Balance sheet finance relates to projects that are being funded by the developer's equity, with well-capitalized developers able to raise corporate debt to provide funding. Project risks are borne directly by the developer, and may be mitigated by bringing in other partners to share these risks.

Offshore wind projects and tidal and wave projects of a size likely to have commercial potential require a large amount of capital investment; as a result, balance sheet finance is only viable for large organizations, particularly utility and oil and gas companies. To date, due to the risks experienced by all offshore energy technologies in relation to project costs, equipment supply, construction and long-term revenue streams, much of the investment in offshore wind energy has been by necessity balance sheet financed. In addition, the restriction in the availability of all project finance during recent years has affected the availability of project finance to offshore wind projects.

For example, the Greater Gabbard 500MW wind farm is currently being constructed as a joint venture between Scottish and Southern Energy and RWE npower renewables as an on-balance-sheet project. However, the burden of these investment programmes during a period of economic downturn may restrict the availability of balance sheet financing for future projects. Other utility companies involved with balance sheet financing of offshore wind farms include Centrica, DONG energy and E.ON. Once a project has been balance sheet financed, there is the potential for refinancing to release equity investment for other projects. For example, the Boreas wind farm portfolio developed by Centrica was refinanced using non-recourse project finance in 2009, with a syndication of 14 banks involved in the transaction.

For wave and tidal technologies, projects at this stage are likely to require a high degree of balance sheet financing, as the immaturity of the market will discourage lenders from investment. In fact, much of this investment may be considered to be in the area of research and development, and may require government support in order to incentivize and encourage investment. In addition, there may be a requirement to increase the level of revenue support available to these projects to enable investment of any kind. The scale of the larger tidal projects that may be required in order to benefit from economies of scale inherent in the technologies may be beyond the balance sheets of even the largest potential investors, and require some form of government support.

7.2.1 Project finance

Non-recourse or limited recourse project financing requires the developer to establish a special purpose vehicle (SPV)[1] for the project. Depending upon the level of debt financing, the lenders will carry the majority of project risk; as a result, the lenders will typically go through an exhaustive process of due diligence to ensure their exposure is appropriately mitigated. Developers typically seek project financing with debt provision from lenders in the order of 60 to 80 per cent of the total investment requirement for the project, and will seek to maximize the percentage of debt in the project to minimize the required equity, which will depend upon the perceived risk of the project and the risk appetite of the market.

Lenders will typically be headed by a mandated lead arranger or arrangers, who will take the leading role in securing financing for the project. Other banks will also participate and distribute the risk for the project. The usual providers of senior debt for project finance, including renewables projects, are the large international commercial banks who have the expertise to assess project finance risks and the appetite to lend to these types of risk. For example, there are approximately 30 to 40 banks generally active in the UK project financing market, of which 10 to 15 could be classified as market leaders able and willing to lead and arrange a transaction and to mobilize other banks to follow. Particular banks specialize in certain types of transactions, often drawing on expertise that they have developed elsewhere in the world. Financial advisers assist in targeting the correct banks and in advising on the specific way to approach them.

Subordinated debt (often called mezzanine debt) is a layer of financing that comes in priority of payment after senior debt and before equity. Subordinated debt plays a role in bridging the gap between what the senior lenders are prepared to provide and how much equity is available for a project. Providers of subordinated debt for renewables projects are typically banks, investors, equipment suppliers and contractors, who would normally tie the provision of financing to an equipment supply or construction contract.

Given the relative maturity of the offshore wind industry, lenders have become more willing to lend under project finance arrangements. The 120MW Q7 wind farm off the cost of the Netherlands was the first offshore farm to be funded through limited recourse financing; since then a number of further projects have progressed with project financing. However, the availability of project finance remains a key constraint to the funding of offshore wind projects.

Financing terms will, to a great extent, be dictated by the project risks, their distribution between the project participants and the mitigation measures in place. Repayment provisions are usually a function of the project economics, and lenders will require full repayment of their loans well within the period of the major contracts – in particular, the power purchase agreement. Lenders will normally be prepared to see their repayments tailored specifically to the cash flow profile of a project. A typical repayment term for a renewables project would be approximately ten years from start-up of the project, depending upon the term of the major contracts. The financing terms will be reflected in a project financial model, which allows stress testing of key assumptions and the impact of these upon the ability of the project to meet its debt repayments to be assessed. This may result in the imposition of additional requirements for reserve accounts or contractor guarantees to be imposed to protect the lenders' investment.

For tidal and wave energy projects, the immaturity of the industry and the uncertainty over project viability make project finance a difficult source of funding at the present time. It is likely that considerable progress in proving the technology and possibly changes in the incentives available through government funding would be required before lenders are willing to provide project finance to these projects. For the very large tidal projects that are likely to be required in order for economies of scale to be realized, the challenges for raising project finance are considerable and may require significant government intervention.

Case study Q7 Wind Farm

The Q7 Wind Farm, now known as the Princess Amalia Wind Farm, was the first offshore wind farm to be project financed and commenced operations in 2008. It is located 23km off the coast of the Netherlands and has a total capacity of 120MW. The project was developed by ENECO, Econcern and Energy Investment Holding, and financed on a non-recourse basis by lenders led by Dexia, Rabobank and BNP Paribas. In addition to the sponsors' equity investment, the debt financing of €189 million was spread over an 11-year project finance facility, with a €30 million standby facility to cover contingencies. The Danish export credit agency Eksport Kredit Fonden also participated in the financing to support the export of the Vestas turbines.

Thornton Bank Wind Farm is 28km off the Belgian coast and has a total capacity of 30MW, planned to increase to 300MW. It is the second offshore wind farm to be project financed. The project was developed by C-Power, which is a partnership of the maritime engineering specialist DEME, Ecotech Finance, SOCOFE, Nuhma and EDF Energies Nouvelles. Dexia Bank was the mandated lead arranger, with Rabobank as mezzanine lender.

Bligh Bank Offshore Wind Farm is 46km from the Belgian port of Zeebrugge, and financial closure for the first stage of the project was reached on 27 July 2009. The first stage of the project has a capacity of 165MW and is intended to reach 330MW, with subsequent stages. It is the third offshore wind farm to be project financed. The project suffered a number of setbacks, including the bankruptcy of the developer Econcern during the development process. The mandated lead arrangers for the project are ASN Bank, Dexia Bank Belgium and Dexia Credit Local and Rabobank International, with mezzanine lenders Rabobank and Maatschappij Vlaanderen. The European Investment Bank is also providing €300 million of project finance for the project. Table 7.1 presents the key features of the three aforementioned project finance offshore wind projects (Project Boreas, which was project financed after construction, is also included for comparison purposes). There are a number of key observations that can be made, such as the small number of construction contracts, long-term O&M contracts, high contingency provisions and long-term lender commitment. These observations confirm the findings of this book.

Table 7.1 Project finance: Offshore wind farms – key features

	Q7	C-Power	Belwind	Boreas
Country	The Netherlands	Belgium	Belgium	UK
Regulatory support – investment	Accelerated depreciation (150% of investment)	Grid subsidy (€25 million)	Grid subsidy (€25 million)	n/a

	Q7	C-Power	Belwind	Boreas
Country	The Netherlands	Belgium	Belgium	UK
Regulatory support – production	Green certificates (fixed – 10 years) €97/MWh	Green certificates (fixed – 20 years) €107/MWh	Green certificates (fixed – 20 years) €107/MWh	Green certificates (market – 15 years) £74/MWh Regulated floor
Electricity sales	Market	Market	Market	Market
Power purchase agreement (PPA)	Yes	Yes	Yes	Yes
Floor/fixed price	Yes	Yes	Yes	Yes
Size	120 MW	30MW	165MW	194MW
Turbine	Vestas V80	Repower 5M	Vestas V90	Siemens 3.6MW
Distance from shore	23km	27km	46km	5km
Water depth	19–24m	15–24m	15–37m	6–11m
Net P50 capacity factor	39%	43% (6 turbines)	37%	38%
Balance of plant (BOP) contractor	Van Oord	DEME/ Fabricom	Van Oord	n/a
Number of contracts	2	3	2	6
Operator	Vestas	Repower	Vestas	Siemens
Initial O&M contract	5 years + 5	10 years	5 years + 5	5 years

	Q7	C-Power	Belwind	Boreas
Country	The Netherlands	Belgium	Belgium	UK
Base case budget	€383 million	€153 million	€619 million	£460 million
(in € million/ MW)	3.2	5.1	3.75	2.4–3.0 (estimate)
Mezzanine, subsidies and early generation	145 million 38%	32 million 21%	89 million 14%	–
Senior debt (base)	€188 million	€95 million	€426 million	£340 million
(as a percentage of base investment)	49%	62%	70%	74%
Contingent budget	€60 million	€16 million	€80 million	Not relevant
(as a percentage of base investment)	16%	10%	13%	n/a
Contingent debt	30 million (50%)	11 million (70%)	56 million (70%)	Not relevant
Contingent equity	30 million	5 million	24 million	Not relevant
Sponsor equity	€50 million	€26 million	€104 million	£120 million
(as a percentage of base case)	13%	17%	17%	26%
Debt – equity limit	50:50	70:30	70:30	n/a

	Q7	C-Power	Belwind	Boreas
Country	The Netherlands	Belgium	Belgium	UK
Maturity	11 years	16.5 years	16.5 years	15 years
Margins (commercial)	125–225pb	110–190bp	300–350bp	300–380bp
'Commercial' debt	148 million/218 million	116 million/116 million	120 million/482 million	340 million/340 million
(as a percentage of senior debt)	68%	100%	25%	100%
International Finance Institution / European Credit Agency (IFI/ECA)	Eksport Kredit Fonden (EKF)	–	EKF & EIB	–
Arranging period	03/05–10/06	06/06–05/07	04/08–07/09	09/08–10/09
Syndication	3 banks	3 banks	n/a	14 banks

Source: Guillet (2009)

7.3 Conclusions

Eventually, the risk profile of a project will reach a stage where implementation will be seriously considered by its owners. At this stage, a number of financing options are available, which for projects developed in the private sector are primarily balance sheet and project finance. Each option has its benefits, with the characteristics of individual projects and their sponsoring organizations typically dictating which one is best for the particular project. Balance sheet finance using debt raised corporately is cheaper, involves fewer parties, and control of the project remains firmly with the owner; it is, however, capital intensive and the risk of failure lies entirely with the owner. On the other hand, project finance allows greater leverage from the available funds for sponsors' equity investment; however, it is typically more expensive and complex and an element of control over the project is afforded to the lenders.

Wave and tidal stream projects have not, to date, been project financed. With technologies still pretty much at the prototype/pilot stage, they are seen as containing large amounts of technology and performance risks. Funding for deployment to date has tended to tap venture capital or public-sector development support sources. A project finance model may emerge in the future once the technologies have been de-risked.

Note

1 A special purpose vehicle (SPV) is a separate legal entity created for a specific purpose – in this case, the ownership of an offshore renewable generation plant.

8

Technical Barriers and Mitigation Measures

A number of technical barriers and challenges are faced by offshore renewable technologies. This chapter covers the main challenges shared by all marine renewable energy technologies, such as technology and design optimization, reliability, installation and decommissioning, operation and maintenance, grid connection, and integration. Finally, we discuss the specific barriers and challenges facing offshore wind, and wave and tidal technologies, respectively.

8.1 Barriers common to all offshore renewable technologies

8.1.1 Technology and design optimization

The respective historical development of offshore wind turbine generator (WTG) technologies, on the one hand, and of wave and tidal technologies, on the other, differ in their origin and optimization. Offshore wind turbine technologies and designs have initially evolved from the converged, dominant 'Danish' design for onshore wind technologies – namely, horizontal axis, three-bladed turbines. The first offshore wind turbines were merely 'marinized' versions of existing onshore WTGs. Furthermore, the first offshore WTG designs and development have been performed by established (onshore) wind turbine technology developers such as Vestas and Siemens, who have at their disposal the technical and financial strengths to pursue these new developments in parallel to their on-going commercial activities. It is only during the last decade that new offshore wind turbine technology developers (Multibrid/Areva, Bard) have gradually entered the market aiming to offer WTGs purposely designed for offshore wind applications, while long-established manufacturers (REpower, Vestas, Siemens) have now started designing WTGs, targeting the offshore market. The trend is now continuing and possibly even accelerating with more radical designs being considered, such as floating offshore wind turbines, two-bladed designs and even vertical axis concepts (see section 3.1 in Chapter 3).

New technical challenges and barriers have also emerged as a consequence of offshore wind development now being considered further offshore and in deeper water depths (> 30m). In response to these challenges, a number of new design and technology solutions have been suggested and investigated as potential solutions. These include new foundation design and concepts for deeper waters (Carbon Trust, 2010), special purpose vessels and revolutionary installation methods.

Regarding wave and tidal technologies, those currently under development are very diverse, in part as a result of the large number of generally small and cash-strapped technology developers pursuing the opportunity provided by a growing international interest in sustainable energy production. Wave and tidal technology designs are likely to gradually converge as a combined result of technical optimization (emergence of 'better' economically viable design options) and commercial attrition (bankruptcy/acquisitions). Although, on the one hand, a large number of wave and tidal designs can be seen as a positive aspect for the industry, which encourages competition and therefore the emergence of viable designs, it is seen externally by many utilities and investors as immature, leading to a dilution of public support measures and incentives.

8.1.2 Reliability

Due to their offshore location, it is crucial for all offshore renewable technologies to be reliable, have maximum operational availability and have minimum planned or unplanned maintenance downtime. The total loss (downtime) or partial reduction in the power generation capacity of an offshore renewable project, due to the component failure, is likely to be exacerbated as a consequence of reduced device accessibility in an offshore environment. Accessibility can be severely limited as a consequence of poor weather (limited wind, wave or tidal windows for vessel or helicopter access), or difficulties and costs in mobilizing suitable vessels at short notice in the case of unplanned maintenance operations. Typical technological solutions and methodologies used by technology developers to improve reliability and reduce downtime include detailed reliability analysis, improved design, the inclusion of reliable sub-components and systems, the inclusion of redundancies, as well as condition monitoring of the plant and equipment.

Although offshore wind technologies and projects have been shown to work effectively, early industry experience has highlighted the fact that reliability problems have led to costly repair and refurbishment interventions.

As highlighted previously, wave and tidal technologies are currently at a lower level of technological maturity compared to offshore wind or other mature electricity generation technologies. Wave and tidal technology developers often highlight the use of proven reliable components in their designs; but only a handful of full-scale devices have been able to openly demonstrate high levels of reliability in open sea. This is not surprising given the level of maturity that this industry has reached overall. Most wave and tidal devices are still to be considered as prototypes or pre-commercial devices. Consequently, they cannot be expected to have achieved a similar level of reliability to more mature technologies, which benefit from experience gained in the field over many years of operation.

8.1.3 Installation and decommissioning

The installation and, in due time, the decommissioning of offshore renewable technologies are characterized by many technical challenges. Although some good experience has been gained in the deployment of offshore wind turbines in shallow waters, the offshore wind industry faces new technical and logistical challenges as wind farm projects are being developed in deeper waters, further offshore, and in a greater variety of seabed conditions. Many project developers and offshore wind, wave and tidal technology developers are considering new installation methodologies. These new methodologies often use innovative and specialized installation and support vessels in order to shorten the construction timescales, increase the weather window envelopes and range of water depths for conducting the work, or improve the project logistics and reduce the number of rotations required between development sites and ports.

Technical barriers are greater for the wave and tidal energy industry where few developers have gained installation or operational experience. Tidal energy technology developers face the additional challenge of strong tidal currents, usually imposing further restrictions on installation windows, which is often only tolerable in slack waters.

Decommissioning strategies have to be considered, and technically and economically viable solutions adopted, as planning permits frequently include strict conditions requiring offshore project sites to be reinstated at the end of the project lifetime. In some particular cases (piled foundations), transferable solutions are available from the offshore oil and gas industry.

8.1.4 Operation and maintenance

The technical and logistical challenges associated with the operation and maintenance (O&M) of offshore renewable energy technologies constitute another important barrier to the deployment of offshore projects. Operation and maintenance strategies are strongly linked to the reliability and accessibility of the operating assets. New access methods (for the renewable energy industry – not for the oil and gas industry), such as helicopter access or permanent offshore living quarters, are being considered or proposed as part of the O&M strategy for distant offshore wind farms. The method of access varies considerably depending upon the offshore renewable technology and adopted maintenance strategies.

A number of high-profile cases, particularly in the offshore wind industry, have demonstrated the high costs and technical challenges associated with major unplanned maintenance work, which in some instances have required complete offshore wind turbine nacelles to be brought back to shore. Reducing maintenance requirements and optimizing operation and maintenance strategies are at the forefront of considerations of all offshore renewable energy technology developers. A number of research and development (R&D) and industry support programmes are in place at company, national and international level to support the development of innovative solutions, as well as other measures and initiatives to remove or reduce these technical barriers.

8.2 Barriers specific to offshore wind technologies

The main technological barriers facing offshore wind energy technologies and projects have been highlighted in the previous section.

Other technical barriers worth highlighting which are specific to offshore wind energy projects include:

- assessment and understanding of offshore wind energy resource, wakes and turbulence;
- lifting and replacement of heavy components located in the nacelle; and
- personnel access and transfer between wind turbine and vessel or helicopter.

8.3 Barriers specific to wave and tidal technologies

Unlike offshore wind turbine technologies, which extract energy from the wind and are therefore in 'indirect' contact with the marine environment, wave and tidal technologies are, by their nature, extracting energy from the waves and tides and their working elements in 'direct' contact with their surrounding environment. Consequently, wave and tidal technologies are facing a number of specific common challenges, some of which are summarized in this section. The specific experience of a number of wave and tidal technology developers can be further consulted (Beaudoin et al, 2010).

8.3.1 Marine environment

Water ingress

The majority of wave and tidal technologies are designed to prevent water ingress into the device through the use of seals, bearings and pumps in order to protect mechanical, hydraulics and electrical systems and components. The abrasive nature of the marine environment and high pressures experienced all lead to significant technological challenges associated with preventing or controlling water ingress.

Bio-fouling

The potential impact of bio-fouling is another technological challenge faced by wave and tidal technologies. Bio-fouling occurs when marine organisms colonize submerged structures, potentially leading to important build-up of organic material over time. While such build-up may not affect some support or passive structures, it has the potential to lead to reduced efficiencies or failures affecting the operation and efficiency of the devices. Bio-fouling may be reduced on moving parts of offshore renewable energy devices, as continuous movement above 1m per second stops organisms from getting an initial foothold. Another issue caused by bio-fouling is that corrosion underneath it can be made much worse. Anti-fouling techniques and paint are a common solution.

Corrosion

Submerged steel structures will be exposed to corrosion, which they will be required to withstand for the operating lifetime of the devices (usually 20 years) unless some major retrofitting is considered in the intervening period.

Wave and tidal resource assessment and performance prediction

The assessment of wave and tidal energy resource has been briefly highlighted in Chapter 2. Large uncertainties remain in the accurate assessment of wave and tidal resources at a specific project location. When assessing the potential energy extracted by a single or an array, additional uncertainties are introduced in order to account for the devices' power extraction curves (tidal) or matrices (wave), and array effects. Further research and development programmes and initiatives are being conducted in order to improve the understanding of the marine energy resource, devices' performance, and to reduce uncertainties assigned to wave and tidal energy yield estimates.

8.3.2 Survivability

Ensuring survivability is probably one of the biggest challenge facing the wave and tidal industry. Occurrences of total loss and sinking of some prototype marine energy devices in the past, and more recently, have often resulted in the corresponding technology development being abandoned, loss of faith in the company and commercial failures. Understandably, wave and tidal technology developers are therefore particularly careful when deploying prototypes in the open sea and initially selecting sheltered conditions (e.g. offered by marine energy test centres such as the European Marine Energy Centre in Scotland).

In order to 'survive', a marine energy device must be designed to withstand the extreme high load forces that it will experience from the waves and tides (e.g. during storm conditions). Understanding of the load cycles over the device's lifetime is also of key importance. One particular challenge consists of accurately estimating these extreme loads and cycles over the device and project lifetime.

Survivability also includes the requirement for devices to be secured in their deployed location. Some wave and tidal technologies have opted for mooring systems which present advantages such as allowing the device to weathervane and are usually lighter and therefore cheaper than fixed structures and foundations. A number of design options (Harris et al, 2004) have to be considered to achieve the optimal compromise between maintaining position and optimizing resource extraction.

In the particular case of tidal energy projects, the electrical cable infrastructure is also likely to face additional survivability challenges as tidal channels are likely to present a barren, hard rock seabed structure, which will not allow these cables to be buried (as in the case, traditionally, of offshore wind farms) in order to provide them protection against wear and tear, or accidental damages from anchors.

8.4 Mitigation and removal of technology barriers

8.4.1 Research and development

As discussed in section 5.2 of Chapter 5, the levelized cost of energy of offshore renewable energy technologies is still very high compared to onshore wind or conventional thermal solutions. Sustained and significant investments in research, development and deployment (RD&D) are therefore a key requirement in order to reduce capital and operational expenditure associated with these technologies.

The removal of technology barriers is performed through investments in fundamental and applied research and development (R&D) activities in the following categories:

- public level (universities, national research institutions);
- private level (in-house R&D within industry and particularly by technology developers); and
- through public–private partnerships or funding programmes.

The amount and nature of investment provided is heavily dependent upon:

- level of maturity reached by the industry;
- strategic importance given to the industry by the country; and
- strategy used in order to assign these investments.

Significant investments in publicly funded research in wind energy were historically initiated during the late 1970s in a number of countries (for example the US, Denmark, the UK, Germany and the Netherlands) as a result of the oil crises. While, in some countries, governments subsequently reduced or abandoned their support to wind energy, others such as Denmark continued to support their nascent industry at times when wind energy technology manufacturers had limited resources to invest in research and development. Interesting differences and impact upon success can also be observed in the way in which these R&D investments were channelled. For example, Denmark favoured incremental developments in turbine size based on lessons learned from operational machines and close collaboration with manufacturers, while the Netherlands and many other countries experimented with large-scale prototypes led by research laboratories and universities.

As the wind energy industry developed and onshore wind turbines reached commercial maturity, publicly funded R&D reduced while support remained in the form of production support incentives (such as feed-in tariff, tradable certificates or tender processes). Wind turbine manufacturers and their suppliers, like most technology critical businesses, are now investing heavily in research and development activities specific to their technologies, with, for example, Vestas spending some €90 million on R&D in 2009. Current leading offshore wind turbine manufacturers by market share (Vestas, Siemens and REpower) have the financial resource to substantially invest in their own R&D. Many of the other established onshore wind turbine manufacturers have announced their intention to enter the offshore wind market and have either initiated R&D programmes or developed prototypes (e.g. Gamesa and Enercon) or acquired small offshore wind turbines developers (GE acquisition of Scanwind). Other multinational energy or engineering companies have entered the offshore wind market through acquisition (e.g. Areva's acquisition of Multibrid or XEMC acquisition of Darwind). A number of other small (by market capitalization) offshore wind turbine manufacturers or developers (such as Bard or Clipper[1]) are likely to become part of larger ventures in the near future, both in order to maintain or sustain the level of R&D investments required by these technologies, and in order to have the financial strengths expected by the offshore wind market to support projects. As the size and opportunities provided by the offshore wind market expand, an increasing number of technology developers are entering the market and investing in R&D.

Public investments can nevertheless still play a crucial role in supporting maturing industries, such as offshore wind. Public investments can be particularly efficient when used to address issues faced by the whole industry and therefore reduce risks and costs for the whole of the offshore wind energy sector.

Examples of such funding activities include projects or technology demonstration funded by the European Union's framework programmes (currently FP7) or national-funded programmes. Often, requirements linked to financial support will include industry-wide collaboration and public dissemination of results and findings. Examples of such funding initiatives include:

- support to the demonstration of offshore wind projects in deep waters in the UK (Beatrice wind farm) and in Germany (Alpha Ventus);
- POWER, RELIAWIND and other framework programme-funded projects.

A recently announced initiative also worth mentioning is the significant financial support to be provided by the European Investment Bank through the NER300 initiative (EREC 2010). This programme is targeting a number of low-carbon technology demonstration projects across the EU, and includes offshore wind and wave and tidal technologies.

Increasingly, public R&D funding is used as a way to leverage private funding. For example, the Energy Technology Institute (ETI), set up in 2007 in the UK, and investing in a number of offshore wind and marine energy related R&D programmes, is to receive UK£500 million of public funding over ten years, to be matched by an equivalent sum by up to ten large industrial partners (including, so far, Rolls Royce, E.ON, EDF Energy, BP, Shell and Caterpillar) and each contributing UK£50 million over ten years in order to be able to take part and benefit from the outcome of these research programmes (ETI, 2010).

A list of some of the major R&D and other technology support programmes and initiatives identified for the 18 countries considered in this book is presented in Table 8.1.

Table 8.1 Government R&D support programmes available for offshore renewable energy technologies[2]

Country	Programme
Belgium	– Energy Fund - Supported Research (2003): EUR 1m for demonstration projects – Annual Renewable Energy And Energy Conservation RD&D Tender (1999): Yearly EUR 5m tender for renewable project funding
Canada	– Clean Energy Fund (2009-2014): EUR 110m total over 5 years to support renewable, clean energy and smart grid demonstrations – EcoENERGY Technology Initiative (2007-2011): EUR 172m funding for renewable/CCS projects
China	– No specific programme identified
Denmark	– Agreement on Danish Energy Policy (2008-2011): EUR 3.4m/year for wave and solar power research. Agreement to push for EUR 100m – 135m R&D funding – Promotion of Renewable Energy Act (2009): EUR 3.4m funding for small-scale grid connected renewable energy projects

Table 8.1 continued

Country	Programme
Finland	– No specific programme identified
France	– OSEO Innovation For SMEs (2005): EUR 500m in funding, grants and preferential loans for renewable projects
Germany	– Climate Protection Investment From Sale Of Carbon Allowances (2008): Allocated EUR 400m from sales of carbon allowances for investments in low-carbon projects – Fifth Energy Research Programme (2005): Around EUR 130m/year funding for renewable projects
Ireland	– Sustainable Energy Incubator Programme (2007): Small (EUR 10,000) grants for ten renewable generation projects – Renewable Energy R&D (2002): EUR 16m for renewable projects EUR 2m for Ocean Energy Prototype Fund and others as part of Ocean Energy Development Unit (UCC 2009) – SEAI: EUR 16m prototype development fund
Italy	– No specific programme identified
Japan	– Subsidy For R&D For New And Renewable Energy (1997): Around EUR 272 m/year renewable project funding
Netherlands	– Energy Research Strategy (2004): Provides around EUR 150m/year funding for renewable research projects
Norway	– Clean Energy For The Future (2004 – 2013): Funding for research and small scale pilot projects
Portugal	– Wave Energy Pilot Zone (2008): Establishes a Pilot Zone for installation of up to 250 MW wave power off São Pedro de Muel – Grant funding: considered, to be confirmed
Spain	– National Plan For Scientific Research, Development And Technological Innovation (2008-2011): Funds renewable R&D activities
Sweden	– Measures to support wind farms in difficult locations (2008-2012): EUR 39m/year funding to help industry gain experience building wind farms in difficult areas including offshore
Taiwan	– No specific programme identified to date

Country	Programme
UK	– Renewable Energy Strategy (2009): Up to EUR 478m funding key *emerging* renewable technologies e.g. ocean, offshore wind – Energy Technologies Institute (2007): Provides funding and guidance into *emerging* renewable technologies. Will invest EUR 710m over 10 years through both public and private finance – Marine Renewable Deployment Fund (2005): EUR 59m funding for wave and tidal stream projects[3] – Marine Renewable Proving Fund 2009 (MRPF): EUR 26m fund administered by the Carbon Trust – Marine Energy Accelerator, Carbon Trust, EUR 4.1m – Research Councils Energy Programme (2004): Provides finance for energy research at UK universities – Scottish Executive: Marine Energy Fund – Scottish Executive: Saltire Prize (EUR 12m) – Technology Strategy Board (EUR 14m)
USA	– American Recovery And Reinvestment Act (2009): Contains over EUR 60 billion funding for clean energy research development and deployment – Energy Policy Act (2005): Authorises subsidies for renewable technologies – Department of Energy: Grant funding, Wind and Water Power Program.

Source: IEA (2010), Dalton et al (2009), Wind Power Monthly (2010), Mott MacDonald

Private research and development activities are led by technology developers, their suppliers in the supply chain and by utilities. A large number of innovations are currently taking place (e.g. in fields related to offshore installation vessels, foundations, generators, condition monitoring and materials).

It is also worth noting that many public research programmes and funding activities are directed towards the mitigation of non-technical barriers and, in particular, environmental barriers (e.g. by supporting environmental monitoring and research activities).

Although in most cases a clear correlation can be observed between the amount of R&D funding provided and the successful development of an industry, the strategy used to channel this funding and country-specific wider conditions (other barriers, financial support mechanism) also play a key factor in the end result.

The wave and tidal energy industry is currently at a very different stage of technical development compared to the offshore wind sector. A large number of mostly underfunded wave and tidal technologies developers are heavily reliant on public funding support for their sector and technology, although very important private investments have also been secured by the front runners. Governments cannot appear to be backing a particular technology or company ('picking winners') and have therefore designed financial instruments to support technology demonstration, such as Marine Renewable Deployment Funds in the UK or New Zealand, or competitive grant allocations calls such as the WATES I and II schemes in Scotland or similar initiatives in the US by the Department of Energy (see Previsic et al, 2009, for an overview of research, development and demonstration activities in the US). A comprehensive list of grants available for wave energy in eight European countries can be found in Dalton et al (2009). Areas of research requirements for the wave or tidal industry are highlighted, for example, by the work packages defined by the Supergen Marine research programme or by the technology roadmaps (UKERC and University of Edinburgh 2008) and the Wavenet (2003) final report.

8.4.2 Guidelines and standards

The availability of offshore wind-, wave- or tidal-specific guidelines and standards can play a very important role in removing technical barriers by ensuring the use of best practice gained from experience acquired by other projects and industries. Standards also provide a mechanism to streamline the approval process from key stakeholders, such as insurance companies, provided that they have backed the standard. In particular, offshore renewable energy technologies can benefit from the large number of standards and guidelines directly applicable from the offshore oil and gas, shipping, water and marine industries; but due care is needed in interpretation.

Furthermore, a number of organizations (in particular, EMEC, DNV, GL, BSH, ABS, Renewable UK and EPRI) have all taken important initiatives which have led to the development and publication of industry-specific guidelines and best practices. This work on standards is now pursued at international level by organizations such as the International Electrotechnical Commission (IEC), which is driving the development and delivery of new international standards for the offshore wind (IEC 61400 series) and for the wave and tidal industry (IEC 62600 series).

Care must be taken, however, not to block out innovation or competition in the application of standards. Harmonized standards that draw together the requirements of numerous stakeholders can accelerate project development and lower its risk, but can also block or slow down innovation in several ways:

- Developing a harmonized standard is expensive and can take a long time to develop. Changing the standard can also require significant time and resources. This hurdle may be difficult for innovators to overcome, especially if the product is still evolving.

- Harmonized standards often set very specific requirements, which innovations may not meet, because they take a different approach to achieving the same objective.
- The development of harmonized standards can be unduly influenced by stakeholders who have an interest in deterring competitive innovation, such as powerful trade associations or governments seeking to protect their businesses.

In order to accelerate project development, lower risk and allow opportunities for innovation, it is helpful to take advantage of several types of product approval mechanisms in a technology regime:

- Harmonized standards: these incorporate the concerns of key stakeholders.
- Technical approvals: where a harmonized standard does not exist or the technology differs significantly from the standard, a technical approval may be granted. Technical approvals assess the ability of a technology to perform a specified function. They are highly regarded in the construction industry for their ability to be flexible and also to provide a robust level of certification. Technical approvals take the form of European technical approvals and those within the European Union of Agreement, such as the British Board of Agreement.
- If a technology is not prepared for full certification, such as a technology in demonstration phase, a technical assessment may be undertaken to validate that key aspects of technology meet critical requirements for a specific application (e.g. safety tests required for a product trial).

Other ways to facilitate the use of standards and avoid negative impacts upon innovation include the following:

- applying standards internationally, when applicable;
- implementing objective-based standards (i.e. what needs to be achieved or complied with) rather than prescriptive standards (i.e. how to achieve or comply), especially in regulation;
- providing support for testing and verification of innovations in the form of guidance, financial support, standards development and test facilities.

8.4.3 Test centres

The development of test centres providing technology developers with consented infrastructure to deploy and field test their technologies has been considered as a high priority by the wave and tidal energy technology communities and industries. Smaller-scale components may also require laboratory testing to meet standards or demonstrate that they are appropriate, as discussed above.

A number of sea trial test centres are now operational for the benefit of the wave and tidal industry, with many more in development across Europe (Waveplan, 2009) and worldwide.

The situation is quite different for the offshore wind industry as, to our knowledge, no independent 'dedicated' offshore wind test sites currently exist, although some are now being planned in the UK (NAREC in Aberdeen).

Furthermore, a number of now operational small offshore wind farm sites have been developed with the expressed intention of serving as research and development platforms and technology demonstrators' sites. These include the 60MW (12 x 5MW) Alpha Ventus offshore wind farm in Germany (RAVE research project) or the 10MW (2 x 5MW) Beatrice offshore wind farm in the UK. Both projects have benefited from public-sector investment and funding.

8.5 Electrical connection, transmission and grid integration barriers

Historically, the technology of choice for exporting power produced by offshore wind farms into the onshore grid infrastructure has been alternating current (AC) transmission. The preference for AC transmission has been driven by the economics of large electrical plant and the relatively short distances (< 100km) between offshore sites and onshore point of connection. However, AC transmission becomes increasingly less economic as the distances increase from shore, mainly due to losses generated in the AC transmission cables.

With future offshore wind farm projects being sited increasingly further from the onshore point of connection, high voltage direct current (HVDC) technology is now increasingly considered as an economical and technically competitive alternative to AC (EWEA, 2009a, p29). A direct current (DC) system requires an AC/DC converter station both offshore and onshore, with both stations being significantly larger installations. The first offshore wind park connection using HVDC technology was completed in the summer of 2009 by ABB for the 400MW Bard Offshore 1 wind farm in the German EEZ (EWEA, 2010b, p46).

HVDC technology is proven in other applications areas, such as long distance transmission on land and sub-sea, but is new as applied to offshore wind projects and is relatively costly. DC power transmission requires costly power conversion and filtering equipment, in addition to massive substructure to support the offshore DC substation platforms. The breakeven distance for considering DC transmission compared with AC transmission depends upon several factors, such as transmission medium (cable or OH line) and different local aspects. However, project–shore distances in excess of 100km are often considered to be optimized for DC transmission.

In addition to offshore grid technologies, it is also of prime importance to consider the most appropriate topology for the offshore grid which needs to be built in order to integrate the future large-volume intermittent energy generated by offshore wind (and, at a later stage, wave and tidal) farms under development and construction. Figure 8.1 represents the current structure and one of many possible future development scenarios for the North European offshore grid. Other scenarios are also being considered.

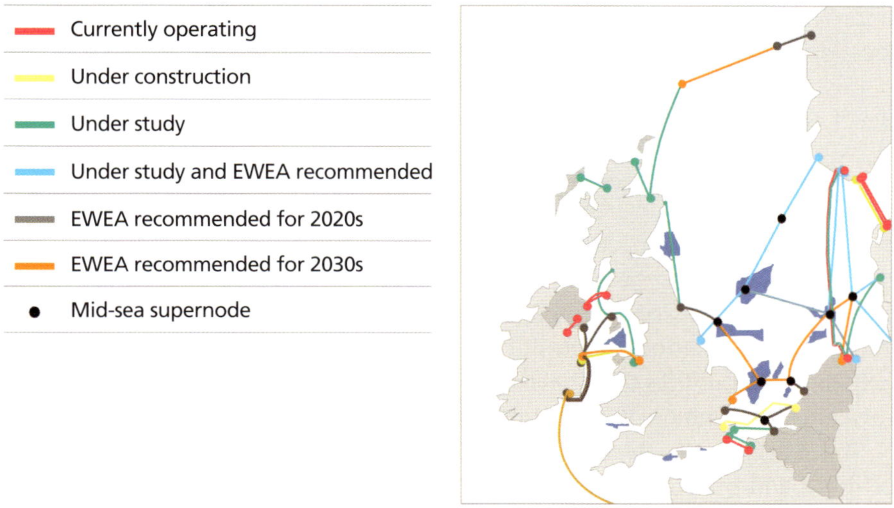

Figure 8.1 Current structure and one possible future scenario for North European offshore grid development
Source: Offshore Valuation Group (2010) and EWEA (2009b)

Current grid infrastructure in individual countries is largely linked to the historical evolution of industry and population (load) centres, centralized electricity generation plants and geography. For some countries or regions (e.g. Portugal or the US and Canadian West Coast), a good match exists between the location of new offshore (wind and wave, in particular) energy resources and the electrical grid infrastructure. Reinforcements to the transmission network will, however, still be required. For other countries, such as Ireland and most of the UK and, in particular, Scotland, a mismatch exists between the resources and the load centres. In order to address any potential geographical mismatch, transmission network reinforcement and upgrades need to be considered. Furthermore, potential mitigation measures such as energy storage and smart grid solutions, which connect and manage the production from offshore renewable energy projects while preserving the grid stability, are significant technical developments that will improve the penetration of renewable energy devices into the existing onshore electrical systems.

At offshore wind farm project level, the electrical energy generated by the individual WTGs is collected and transformed to higher voltages by offshore substations usually mounted on the similar structures (pile/jacket) to the WTGs.

A similar electrical infrastructure will apply to future wave and tidal energy farms, although differences may be explored with respect to offshore substation designs, which could, for example, be placed on the seabed. This technology is being pioneered in the UK at the Wave Hub project to connect up to 20MW of wave energy projects (South West RDA, 2010).

While point-to-point HVDC transmission is proven, this technology is unproven as applied to network infrastructure interconnecting multiple locations.

8.5.1 Grid connection and power distribution

A number of technical barriers to grid connection and distribution of offshore renewable energy have already been highlighted. These include a lack of onshore grid infrastructure, potential weaknesses of coastal grids, and mismatches between geographical locations of offshore developments relative to electrical load centres (DECC 2010a). These technical challenges are, however, in theory largely surmountable if it were not for a number of regulatory barriers within many countries that make changes to existing grid networks and market structures challenging, if not impossible, to achieve in the required timeframe.

The challenge of achieving a grid connection is best described in terms of the offshore and onshore systems, which are clearly different in terms of location and technical considerations. Increasingly, however, the onshore and offshore systems also fall under different regulatory and commercial regimes. This is a crucial consideration for offshore renewable energy development as the time required to obtain the necessary permits and make required modifications to onshore grid networks can be significantly longer than the time required for developing offshore power generation and connections. This makes onshore grid development the rate-limiting step and the ability to resolve the regulatory barriers of onshore distribution the crux of offshore renewable energy deployment.

Onshore grid networks have typically been well established and operate within a set of highly complex regulations, regulatory bodies and commercial arrangements. Offshore networks are essentially starting from scratch, and while permitting and regulatory issues arise, policy-makers have greater freedom to dictate how these connections will be regulated and managed provided that they meet the requirements necessary to obtain seabed rights, including environmental permitting.

These regulatory challenges are further complicated by the chicken and egg scenario that can potentially be established in which financial backing for onshore grid development is lacking without evidence of offshore power development and vice versa, as has occurred in the north and west of Scotland until the recent announcement of the approval of the Beauly to Denny transmission line and further reinforcement in order to allow long-term wave and tidal developments planned in the Pentland Firth to proceed.

Barriers to onshore grid connection

There are a number of regulatory challenges associated with optimizing onshore electricity networks which can present barriers to both onshore and offshore renewable energy development. Obtaining the necessary environmental permits and planning permissions across multiple jurisdictions encompassing numerous stakeholders' interests is an immense challenge which can take many years to work through. Nonetheless, a precedent is in place for achieving this objective for other major infrastructure projects, such as pipelines; so, in principle, this undertaking can be achieved.

However, other systemic issues make the prospect of connecting onshore renewable energy projects substantially more difficult than other infrastructure projects. For the most part, they are rooted in several core issues:

- regionalization of power distribution networks;
- cost allocation;
- lack of competition;
- congestion; and
- prioritization.

Regionalization of grid networks

Power distribution networks are often operated and regulated on a regional basis. From an operational standpoint, the regionalization over ownership and regulation can create pinch points and market conditions that discourage the distribution of renewable energy across regional boundaries. Regional regulators may lack the incentive, authority or mechanism to resolve these regional difficulties in the interest of national policy, and a central regulatory body with the authority to break down regional issues may also be missing.

Regional barriers to grid development: US

The US is particularly vulnerable to regulatory barriers within the grid network. The American Wind Energy Association has identified a lack of fiscal instruments required to raise capital and allocate costs to resolve capacity bottleneck issues as the greatest barrier to large-scale renewable deployment in the country (AWEA 2008)

Lack of competition

A lack of competition in expanding the electricity grid and offering connections to new sources of generation is a core impediment to the integration of new generation. In many markets, transmission system operators are the only provider of grid access and new connections, and a lack of alternative providers can create delays in reinforcing the network and providing connection, resulting in increased congestion.

Historically, in the UK, the 'queue' for access has been managed on a 'first come, first served' basis. As a consequence, less viable projects have, in many cases, blocked those that are further advanced or could be developed faster. It has been the case that capacity has been booked in the queue even when there is no realistic prospect of the project in question being connected by the contracted date. These arrangements have, therefore, led to considerable uncertainty and have acted as a barrier to entry into the energy market.

The building of queues also delays connection of new sources of generation, where in some markets, connection dates of 15 years are offered. Long lead times, a lack of coordination and uncertainty create an adverse environment for developers, which can discourage investors who will shift their capital to other markets. Delays also jeopardize national and supranational policy objectives, such as those mandated by the European Commission. Hence, it is in the interest of regulators and government to have a coordinated approach that ensures commercially viable connections to the grid with firm connection dates that are consistent with the timetable of developers. Delays cause assets to be unutilized or stranded, generating unnecessary costs for consumers and investors alike.

Congestion

Congestion on the onshore electricity grid, particularly for transmission networks, is an impediment to offshore developers. Strictly speaking, it is a technical issue and does not stop the development of new generation assets; but it creates a market environment that creates significantly higher costs for consumers and can act as a barrier to development, which must be taken into account in policy development.

If elements of the onshore network are at capacity, creating additional sources of generation offshore may result in an excess production of electricity that the grid cannot deliver to sources of demand. This is, for example, the case in Germany, where nuclear and fossil fuel power plants are near the coast, requiring additional transmission lines or reinforcement to be made to accommodate the connection of offshore wind farms.

If new sources of generation are brought online and the network has reached capacity, this means that other sources of generation must be taken offline to accommodate the other (known as 'constraining off'). Some generators have allocated capacity for transmitting their electricity to consumers, and constraining off such generators results in compensation. These costs are recovered by system operators from generators and suppliers, who pass on these costs to consumers in their entirety.

Power purchase agreements

Power purchase agreements (PPAs) are contracts between the electricity generator (in this case, the offshore renewable energy owner) and power purchasers (or the grid). The rates agreed within the PPA and other issues such as prioritization of capacity influence the financial model and the bankability of a project.

There are a number of PPA issues that can delay offshore energy development. The first is a conflict of interest between businesses who own both the grid networks (and purchase power) and generation facilities (such as conventional power plants). In such an arrangement, the addition of renewable energy can be seen as a competing interest to the electrical generation side of the business. Full unbundling and transparency is necessary to ensure that PPAs do not discriminate against renewable power generation on the basis of factors such as grid codes or network charges.

Market factors related to issues such as balancing and the relatively low value (in fiscal terms) of wind energy are also potential issues, especially during periods of high wind, when the energy being generated may not be required.

Barriers to offshore network development

Development of offshore distribution networks to reach shoreline grid connections is subject to seabed rights and permitting requirements as power generation facilities. Additional permits may also be required to allow cabling to cross coastal zones.

While environmental issues may differ between power generation and distribution, a cable connection covers a wider area and potentially crosses more jurisdictions, including international Exclusive Economic Zones (EEZs). Consequently, the permitting processes can be equally, if not more, complex, in the process of taking into account:

- fisheries and mariculture;
- shipping/above- and sub-surface transport links;
- exploration and production of oil and gas;
- development of offshore electricity generation (wind, wave, etc.);
- conservation and marine parks; and
- cables and pipelines.

The degree to which barriers affect onshore grid distribution, such as regionalization and lack of competition, is not necessarily set in stone, but can potentially be influenced by policy independent of onshore networks. There are various approaches to market arrangements for developing and operating offshore transmission assets:

- public/government;
- developers/generators;
- onshore transmission operators; and
- independent commercial third parties.

Examples of different regimes include Belgium and the Netherlands, which have offshore developers operating transmission assets, while Germany and Denmark have their onshore grid operator undertaking that role. Alternatively, market arrangements could have independent commercial third parties owning and operating networks, such as large institutional investors or infrastructure companies. The UK's offshore transmission (OFTO) approach presents a different regime which will grant licences to own and maintain new offshore transmission assets via a competitive tender process.

The pros for the owner to develop, build and operate an offshore grid connection are as follows:

- Development, procurement and implementation activities will be controlled by one organization.
- It is easier to coordinate the project and construction schedules of the grid connection and the wind farm, respectively. It should be appreciated that if the grid connection is delayed and the wind farm is completed on time, then commissioning cannot be completed, no revenues can be generated and effort may be needed to provide a separate power source to the turbines and other infrastructure to ensure that the equipment can be maintained.
- In operation, there will be a strong incentive to ensure that any outages of the grid connection are repaired as quickly as possible as the whole revenue of the wind farm will be lost during the outage.

The cons are:

- One grid connection is needed for each wind farm with limited opportunity for adjacent wind farms to share a connection.
- The costs are likely to be higher on a country-wide basis.

The pros for the grid connection and wind farm ownership and operation to be separated are as follows:

- It will introduce more competition in the sector, particularly for the UK model where grid connections will be subject to a separate procurement process.
- For countries with a large concentration of wind farms in a particular area, separating the grid connections will allow more flexibility in system planning

(e.g. allowing more than one farm to use a single connection). In the longer term, separating the grid connection will allow the planning of an offshore network.

- Savings occur through refinancing. Presumably, these arise because the operating phase risks of the electrical connection have a very different profile than those of the wind farm. The electrical connection carries low risk (no market risk and very little availability risk), whereas the wind farm carries electricity market risk and much higher availability risk. Arguably, therefore, if the connection is financed with the wind farm, the cost of debt for the connection part of the financing reflects that additional risk.

The cons are:

- These are largely the opposite of the pros for owner-developed systems.
- In particular, the revenue stream related to owning and operating a grid connection is very small compared to the revenues from a wind farm. Therefore, the grid operator may not be as incentivized as in the case of an owner who also owns and operates the grid connection to make repairs quickly. A wind farm operator may have no financial recourse to the grid operator if a repair is delayed. Such a risk is not typically insurable, although this should be considered on a case-by-case basis by an insurance adviser.

For a single-point connection, even if the endgame is for it to form part of a wider network at some future time, there are good risk reduction reasons for the wind developer to take it on. At the moment, projects financed on the market are likely to be unable to raise finance without this.

The other option of having the grid company build the offshore network (as, for example, in Germany) presents advantages as discussed above; but the network connection being late is very discouraging and likely to result in difficulties with investor confidence in the future and/or large liability commitments for grid operators.

Taking a longer-term strategic view, should genuine offshore networks evolve in the future, there is an excellent case for these to be separately owned as they could offer the possibilities for many wind and marine projects to share networks, bringing scale economies and greater reliability, potentially also allowing wind projects to export energy to more than one country and enabling energy interchange between countries. However, such networks would need to be HVDC and the technology does not yet exist to support HVDC networks of this complexity (although it could be developed).

The decision regarding the most suitable approach to follow needs to be taken to suit the particular conditions in a country and the political landscape. In the EU, there might be additional requirements for ownership separation of offshore networks from offshore energy projects, as is required onshore.

8.6 Mitigation and removal of grid connection barriers

The development of offshore renewable generation is reliant on timely and adequate access to the electricity grid. However this can be undermined by:

- lack of transmission capacity;
- lack of prioritization by the transmission system operator (TSO) or bodies with similar functions;
- permitting delays;
- regionally based frameworks of grid infrastructure, regulation and markets; and
- deadlock between TSOs and the developer in which both parties are waiting for the other to develop infrastructure before investing.

8.6.1 Approaches to offshore transmission

There are various approaches to market arrangements for developing and operating offshore transmission assets:

- public/government;
- developers/generators;
- onshore transmission operators; and
- independent commercial third parties.

Belgium and the Netherlands have offshore developers operating offshore network assets, while Germany and Denmark have their onshore grid operator undertaking that role. Alternatively, market arrangements could have independent commercial third parties owning and operating networks, such as large institutional investors or infrastructure companies.

8.6.2 Approaches to onshore transmission

If elements of the onshore network are operating at near full capacity, creating additional sources of generation offshore may result in an excess production of electricity that the grid cannot deliver to load centres.

Addressing the lack of capacity in the network infrastructure can present a number of investment, market and regulatory and permitting challenges outlined above. Options for resolving some of these underlying challenges are outlined in Table 8.2.

If offshore generation is brought on line without adequate network capacity, some sources of generation must be taken offline to accommodate the additional power, a process referred to as 'constraining off.' Constraining off incurs a cost to compensate the generator that has reduced power output. These costs are recovered by system operators from generators and suppliers, who pass on these costs to consumers in their entirety.

Table 8.2 Transmission network options

Access Regime	Features
Invest and then connect	Generators begin exporting only after completion of wider improvements to transmission. No targeting of constraint costs.*
Connect and manage constraint costs and network improvements	New generators may begin exporting at completion of enabling works. Constraint costs incurred and passed on to consumers.
Hybrid solutions	Hybrid solutions can be developed to allocate some of the new constraint costs to new generators due to their accelerated connection.

Note: No targeting of constraint costs means that new connection of generators will continue to take place despite the increased constraint costs arising as a consequence. Targeting of constraint costs means that the objective is to reduce constraint costs (e.g. with network reinforcement).

Constraining is an economic issue that must be addressed in the context of each individual network system. Constraining generators is recommendable if such constraints are rare and lost volumes of energy are low, compared to the cost of new grid infrastructure which would be required to accommodate them. If this is not the case, these constraints may only be acceptable for short timescales (e.g. pending planned reinforcements to the grid network). In order to assess constraining costs, it is necessary to perform a detailed cost-benefit analysis of carbon emissions forgone, the cost of such carbon, compensation, the cost of network reinforcement, and the demand and price of electricity to arrive at a net present value.

Offshore generators may consider a number of options for addressing constraints in transmission networks. Options for improving transmission access have been considered in the UK (see DECC 2010a).

A number of technical solutions may also help to facilitate grid connections, including:

- use of HVDC connection to reduce grid disturbance;
- application of dynamic ratings[4] on grid network to accommodate short-term overload in grid ratings; and
- use of high voltage DC to transit power over a distance to a location where grid connection can be made.

Government-backed investment guarantees have also been suggested as a means to get past the potential deadlock between onshore and offshore required investments in infrastructure, and to remove a 'chicken and egg' scenario in which onshore investments are not occurring unless offshore development proceeds and vice versa.

Options for resolving some of these underlying challenges to grid connection are outlined in Table 8.3.

Table 8.3 Summary of potential mitigation measures for barriers to grid connections

Barriers	Potential Mitigation Measures	Examples
Lack of grid transmission capacity	Paying 'constraining-off' fees for shutting down other generators to create capacity	Paying constraining-off fees (UK)
Lack of prioritization for developing grid capacity for renewable energy as opposed to other priorities	Full unbundling, increase competition, change protocols or regulations to remove projects from the waiting list for transmission capacity if they have a low probability of achieving success within a set timeframe	Unbundling requirements: EU mandatory by 2011 Increased competition (UK)
Allocation of deep grid connection costs[5]	Regulation requiring grid operators to bear costs, including offshore losses	German Infrastructure Planning Acceleration Bill 2006 (POWER 2007)
Regionally based framework of grid infrastructure, regulation and markets	Increase competition National authority	
Deadlock between transmission system operators (TSOs) and developers	Government-backed investment guarantees Allocating investment risk across national parties	Investment guarantees (UK) Allocating investment risk (EC 2005)
Onshore permitting delays	National planning guidelines and authority	National planning guidelines (UK)

Snapshot of UK onshore versus offshore transmission markets

The UK government has set ambitious targets for the development of renewable energy over the next decade. By 2020, the government expects that 15 per cent of the UK's energy needs will be met from renewable sources. This is a significant goal given that the British transmission network experiences significant congestion due to the bulk of generation taking place in the north (Scotland), while the bulk of electricity is consumed in the south.

The UK provides an interesting example whereby different market regimes have been established by the regulator to facilitate connection to the grid for new generators: one for onshore and a new regime for offshore, bringing existing 'transitional' and future 'enduring' changes into force.

Onshore problem

The onshore regime has not been able to deliver appropriate solutions to developers of new generation. A lack of competition in offering grid connections, alongside network congestion, has resulted in connection dates of up to 15 years being offered by the National Grid, also partly due to an absence of a system that prioritizes projects. Demands placed on the transmission network to connect the 71GW of new generation capacity require significant network investment. This queue of 71GW is almost as much as the currently installed capacity of Great Britain.

Onshore solution

The UK government undertook a cost-benefit analysis which focused on assessing the different options to allocate constraint costs that result from connecting new generators to the grid, before full reinforcements have been completed. The analysis found that connecting generators to the network and managing the arising constraint costs, while undertaking necessary network reinforcements, comprise the optimum solution for the onshore regime. This is a regulatory solution to alter the structure of the transmission access market, which has enhanced the ability for offshore generators to access the grid. A significant driver of this solution was to accelerate renewable energy development and meet mandated renewable energy targets.

Offshore regime

In view of the problems faced by the onshore regime in the UK, regulators have opted to implement a competitive offshore transmission operator regime, where competitive bids are invited for licensees to design, build and maintain transmission assets above 132kV for 20 years. This has so far been effective in attracting a significant appetite for investment in an environment that promotes greater certainty than the onshore regime. Existing offshore network assets are currently operated by generators, although these will be sold as part of a competitive bid process to ensure that enduring and transitional arrangements are harmonized.

The UK government and regulator are currently undertaking a consultation to allow generators to build offshore transmission assets before transferring ownership via a competitive tender process. This reduces a significant contractual interface risk between offshore transmission and power generation developers.

8.6.3 Power market mitigation measures

Power markets may present a number of hurdles to offshore renewable energy, including:

- inability to obtain long-term power purchase agreements, in the absence of feed-in tariffs;
- low prices when wind energy is in surplus – for example, under high wind conditions, especially if combined with low demand (night time);
- balancing challenges: the variability of the output affects the value of the power as other generators must compensate for this variability, for which there is a cost; and
- concentration of market power by a small group of large suppliers, which makes it difficult for independent generators/technology developers to access the market without going via the major players.

Potential mitigation measures are outlined in Table 8.4.

Table 8.4 Summary of potential mitigation measures for power market challenges

Barriers	Potential Mitigation Measures	Examples
Lack of long-term power purchase agreements (PPAs)	Use financial hedging instruments Feed-in tariffs	Many feed-in tariffs in place for offshore renewable energy technologies
Low prices in high wind periods	Use financial hedging instruments Feed-in tariffs	
Balancing challenges	Improve short-term wind generation forecasting Reduce time between gate closure and real time trading Use services of an aggregator (which pools risks across other parties) Combine pooling regions Storage	Better forecasting tools Reduce recording period between closure and trading (State of California) Geographically dispersed portfolios
Market concentration	Feed-in tariffs	

8.7 Conclusions

The main technical challenges and barriers shared by all marine renewable energy technologies include technology and design optimization, reliability, installation and decommissioning, operation and maintenance, grid connection, and integration. Considerable investments will be required in onshore and offshore grid infrastructure in order to accommodate the large expected expansion in variable generation capacity from offshore renewable energy projects. In some parts of the world, the optimal topology of this expansion needs to be considered at a supranational rather than national level. Technical barriers are surmountable, but usually affect the cost of offshore renewable energy project and technologies.

Notes

1 Clipper has been purchased by United Technologies (UTC), as announced on 20 October 2010 in the press.
2 All values converted to Euros. See Appendix D for exchange rates used.
3 MRDP funding has not been accessed so far and is likely not to be available from April 2011.
4 Dynamic Rating represents the ability to run a transformer above its capacity for a limited time, safely, and without significant amortization.
5 When new connections take place, their costs must be allocated (somewhere, generally to consumers in liberalised markets). Shallow costs are for allocating the cost of immediate infrastructure necessary to facilitate connection – transformers, etc. Deep costs are for the wider network reinforcement that may need to take place in order to efficiently and safely connect the new generator, i.e. beyond the immediate infrastructure associated with the new generator.

9

Non-Technical Barriers and Mitigation Measures

9.1 Introduction

Offshore renewable energy technologies and projects face a large number of non-technical barriers and challenges to their development and deployment. In this chapter, we provide a description and discussion of the major general barriers and challenges linked to environmental, safety, regulatory and licensing, competing uses, skills, supply chain and infrastructures, and financial issues. Reference is also made to the IEA–RETD RENBAR project and report (IEA, 2010), which focuses on non-technical barriers (specifically on environmental, administrative and social acceptability barriers) and best practices for a number of renewable energy technologies, including offshore wind, wave and tidal. Where applicable, country- and technology-specific barriers are highlighted and case studies presented. After analysing each type of barrier, this chapter goes on to describe the associated mitigation measures that can help to remove those barriers.

9.2 Environmental barriers

9.2.1 Introduction

This chapter provides an assessment of the non-technical environmental barriers to the development of offshore marine renewable energy projects. The nature of environmental impacts is such that similar issues can be expected to arise in all countries seeking to exploit offshore marine renewable energy resources and these types of issues can therefore be viewed as being globally relevant. Where applicable, country-specific barriers and issues are highlighted and described, and case studies are provided. Reference is also made to the RENBAR project and report (IEA, 2010). A large volume of evidence and operational experience has been learned from the Danish Environmental Monitoring Programme

(DEMP) for large-scale offshore wind parks and the current follow-up programme (ENS, 2010). The programme's scientific credibility has been supported by the International Advisory Panel of Experts on Marine Ecology and is internationally widely acknowledged for its methodology development and conclusions.

9.2.2 Overview of environmental barriers to development

The world's marine areas represent a potentially huge, and currently largely unexploited, resource for renewable energy generation to service the world's increasing need for a source of clean reliable power. During recent decades our longstanding relationship with our marine environment has experienced a fundamental shift in many areas of society, and this has entailed a move away from a purely exploitative relationship to one where increasing emphasis is placed on acknowledgement and protection of marine resources.

This shift has gained a particular relevance in the most highly developed and technologically advanced countries, where increased levels of education and technological development have provided the majority of the population a lifestyle which is increasingly less dependent upon direct exploitation of natural resources, including the marine environment. In these societies, the intrinsic value of 'nature' and natural areas is often very highly prized and developments which may affect this can prove highly controversial. This, in turn, has led over time to more demanding legal and permitting requirements to be placed on offshore project developments across industries, including oil and gas, shipping, aquaculture and, more recently, offshore renewable energy. Precautionary principles are also frequently applied to new offshore technologies as a direct result of the lack of knowledge and operational experience gained on their environmental impact.

With regard to the development of renewable energy projects in marine areas, those environmental issues which are most likely to represent significant barriers can be summarized as follows:

- impacts upon marine ecology;
- impacts upon birds;
- landscape and visual changes;
- alteration of existing marine currents or wave regime, and seabed sedimentary processes;
- interaction with historic/cultural assets;
- change of use and conflicts for marine areas;
- impacts upon marine safety and communication systems; and
- interaction with marine and terrestrial protected areas.

The following sections consider each of these headline issues in more detail and those impacts which may present an increased barrier to certain technologies and not to others.

It is worth noting that as the scope of this work is limited to barriers and harmful impacts, the positive impacts of offshore renewable energy technologies (such as, for example, local and national economic development opportunities, and contribution to national and international carbon reduction or renewable energy production targets) are not specifically highlighted.

9.2.3 Marine ecology

Greater scientific understanding of the marine environment has brought an increased focus on marine ecology and the impact of human development upon it. Adverse impacts upon marine ecological receptors have the potential to generate significant opposition from marine industries, conservation bodies and the general public, leading to the refusal and cancellation of projects. In respect of this, the main potential ecological barriers to marine renewable projects are highlighted below for broad categories of receptors. It is, however, important to stress that there are significant variations in impact for the numerous species within the following categories.

Fish

Fish represent one of the most visible receptors which could be affected by the development of renewable energy projects in the marine environment. Potential impacts upon fish extend to both direct impacts upon individuals or populations and indirect impacts as a result of changes to their habitat.

With regard to fish, the most significant impacts are liable to arise as a result of noise and vibration during the construction and operation of generation assets. As such, all types of renewable generators have the potential to affect fish populations, although certain high noise activities such as foundation piling have a higher potential to cause adverse effects. Fish use sound for a number of functions, including communication, avoiding predators and hunting, and can be very sensitive to noise impacts and detect man-made noise considerable distances from the source.

Where fish exhibit avoidance behaviour and leave an area, this can have far-reaching consequences on the remaining ecology and also on human activities such as commercial fishing.

Electromagnetic fields (EMFs) from operating equipment and buried sub-sea cabling are another aspect of marine renewable energy development which is currently being researched with regard to potential impacts upon fish. Electro-sensitive species of fish (including sharks, skate and ray) hunt by detecting weak bioelectric fields emanating from their prey and following these to their source. There is growing evidence that the electromagnetic fields generated by marine generation infrastructure can replicate these weak bioelectric fields and become the focus of electro-sensitive species; the longer-term impacts of this

Recent research on impacts of noise upon fish

Research into the effects of offshore wind farm noise on marine mammals and fish in the North Sea, published by the UK charity COWRIE (Collaborative Offshore Wind Research into the Environment), in 2006 states:

> Cod and herring will be able to perceive piling noise at large distances, perhaps up to 80km from the sound source. Dab and salmon might detect pile-driving pulses also at considerable distances from the source. However, since both species are predominantly sensitive for particle motion and not pressure, the detection radius cannot be defined yet. Behavioural effects, like avoidance and flight reactions, alarm response, and changes of shoaling behaviour are possible due to piling noise. The spatial extension of the zone of responsiveness cannot be calculated, as the available threshold levels vary greatly. The zone of potential masking might in some cases coincide with the zone of audibility. Also physical effects, like internal or external injuries or deafness, and even death may happen in close proximity to pile-driving.
>
> Operational noise of wind turbines will be detectable up to a distance of approximately 4km for cod and herring, and probably up to 1km for dab and salmon. Within this zone, masking of intra-specific communication is also possible. Behavioural and/or physiological (stress) effects are possible due to operational wind farm noise. However, they should be restricted to very close ranges.

The COWRIE study focused on four species of fish: dab, Atlantic salmon, Atlantic cod and Atlantic herring, whose hearing capabilities are well known and represent different levels of fish auditory capability.

Source: http://www.offshorewindfarms.co.uk/Assets/BIOLAReport06072006FINAL.pdf

behaviour are not yet fully understood. However, experience gained from the DEMP has concluded that eels, which would be expected to be very sensitive to electromagnetic fields, were not affected adversely.

Proven mitigation techniques to reduce underwater acoustic impacts exist and can be used to protect fish. These include the use of sonic deterrents (commonly referred to as 'pingers') and the masking of potential sound emitters, such as piles. In addition, ramp-up procedures for pile-driving operations can prove very effective by allowing fish time to vacate the working area before large impacts occur. It is also often possible to minimize potential construction impacts uon fish population by avoiding works in important areas such as nursery areas during key periods of the year. Technologies which are placed in high-energy environments, such as wave generators, may benefit from increased baseline noise levels acting to mask generator emissions.

The effects of EMF can be reduced through increased shielding of emitters and increasing burial depths of cabling.

A number of research programmes (DEMP, COWRIE, RAVE) are being funded in order to gain a better understanding of these impacts and to support the development of mitigation measures.

It is also worth noting the positive reef effect created by offshore wind farms by providing refuge due to limitations on fishery in the parks, as well as new structures for the development of marine life.

Marine mammals

Marine mammals represent one of the highest profile ecological receptors which could provide a barrier to the widespread implementation of offshore renewable energy technologies. During recent years, many marine species such as whales, dolphins, turtles, manatees and seals have enjoyed increasing national and international protection and represent highly visible symbols of marine conservation.

Noise impacts can be of particular importance in the marine environment due to the transmissibility of sound through water. Studies have shown that marine mammals can be very susceptible to acoustic impacts, with hearing frequently being one of the key senses used by marine mammals for communication and hunting. All renewable energy technologies will introduce new sources of noise into the environment; however, the impacts may be very different depending upon the specifics of each project and the receptor.

Large acoustic impacts, such as those that may be experienced from pile driving or rock blasting, can create pressure waves which may cause permanent physical damage to, or even kill, marine mammals in close proximity to the work site. As such, those technologies which require significant foundation construction works (e.g. piled wind farms) can be more susceptible to causing this kind of impact than those which do not (e.g. anchored wave generators). Operational noise impacts are likely to prove less of an issue with regard to marine mammals as the magnitude of noise emissions is liable to be smaller, although it is possible that certain species or individuals may exhibit avoidance behaviour.

As with acoustic impacts upon fish, proven mitigation techniques to protect marine mammals from large acoustic impacts are commonly used in the offshore construction industry and include the use of sonic deterrents and ramp-up procedures for pile-driving operations. Less common techniques being considered include the masking of potential sound emitters, such as piles, by using sound-deadening materials or bubble curtains. Again, environmental factors can help to mitigate acoustic impacts, and technologies which are placed in high-energy environments may benefit from increased baseline noise levels acting to mask acoustic emissions. It is also often possible to minimize potential construction impacts upon marine mammal population by avoiding works during key periods of the year, such as breeding periods, where projects are being developed in important habitat areas.

With particular reference to tidal energy assets, marine mammal collision risk is acknowledged to be another potentially significant, though currently not well-explored, issue. Factors such as turbine blade tip velocities, water motion, mammal swim velocity and underwater visibility can combine to make collision a very real risk. Indeed, collisions between sea mammals and ships are well documented. Collision risks for other types of renewable energy devices such as wave generators are not likely to be significant; however, entanglement with mooring lines or collisions with surface devices in rough seas are possible. However, results from monitoring programmes performed following the installation of the Marine Current Turbine 1.2MW Seagen device in Northern Ireland or from the Verdant RITE project in the US have so far concluded that marine mammals have avoided the turbines and these have not had any adverse environmental impact.

Indirect impacts upon marine mammals can extend to changes in mammal distribution or population structures as a result of disruptions to previously existing food supplies or habitats.

Benthic receptors

Benthic organisms can be subject to a number of potential impacts from marine developments and will, to some extent, be affected by all types of generation device. The most common impacts upon benthic receptors are direct fatalities as a result of crushing or burial by foundation structures and indirect impacts as a result of habitat modification. This can include smothering as a result of alterations to existing sediment erosion and deposition patterns and seabed scour from placed structures and heavy mobile components, such as anchor chains.

The introduction of new hard substrate can provide an opportunity for new species to colonize an area where they were not previously found, altering the pre-existing ecological balance. Moreover, the capacity for such introduced hard substrate to act as an artificial reef and promote the growth of colonizing species can also be a consideration during decommissioning. Removal of foundations and other structures which have been colonized and are acting as artificial reefs can in itself be viewed as a negative impact and a barrier to removal.

Belgian experience

Studies conducted by the Belgian Management Unit of the North Sea Mathematical Models (MUMM) published in 2009 highlighted that the first six concrete foundation structures of the C-Power offshore wind farm on the Thornton Bank showed rapid and heavy colonization within the 3.5 months of being placed offshore. Of the colonizing species, four were identified as being non-native invasive species.

Source: Degraer and Brabant (2009)

9.2.4 Birds

With regard to impacts upon birds, these are primarily considered to relate to the development of offshore wind turbines and not wave or tidal generators. However, that is not to say that wave and tidal generators will not have an impact. In periods of low visibility, any generation asset which protrudes above the surface of the water has the potential to pose a collision risk to low-flying birds and there is growing concern in some quarters as to the risk posed to diving birds by submerged assets, an issue which had not hitherto been considered as a danger.

Likely impacts upon birds from wind turbines can be broadly broken down into three main categories:

- collision risks;
- habitat loss; and
- barrier effects.

As with onshore wind farms, bird collisions with turbine structures, including blades and towers, resulting in fatalities and injuries represents one of the major concerns in respect of offshore wind turbines. Much research has been directed towards identifying the significance of this risk and possible mitigation measures to prevent it. Unlike onshore developments, however, the nature of the marine environment is such that fatalities and injuries are very difficult to estimate as the bodies of the killed and injured birds do not remain at the site of the incident but are often carried off by current and wave action, and subject to predation or sink.

Habitat loss is another key barrier to widespread implementation of renewable energy devices. Certain bird species are very susceptible to anthropogenic disturbance and can display significant avoidance behaviour, including abandoning areas where such disturbance occurs. This behaviour has been observed at a number of existing offshore wind farms; however, due to the limited time during which the majority of marine wind farms have been operational, the implications of the longer-term impacts of this avoidance behaviour are still being assessed. Nonetheless, displacement of individuals from part of their range has the potential to result in overcrowding and increased pressure on resources in surrounding areas. Where resources are limited or poor, this can lead to increased mortality and declining population sizes.

Potential barrier effects from offshore wind farms are another growing concern amongst legislative and conservation bodies. Where migratory bird species display avoidance behaviour, flocks may deviate from their normal migratory routes to 'go around' turbines, leading to increased energy expenditure and longer flight times (with a corresponding decrease in the condition of the birds). While this may not present a significant risk where one offshore wind farm exists in isolation, the barrier effect can be magnified significantly where

Horns Rev wind farm

Following 15 surveys carried out over the period of 2003 to 2005, research published by the National Environmental Research Institute of the Danish Ministry of the Environment (Petersen et al, 2006) showed that, in terms of red- and black-throated diver populations (*Gavia stellata/arctica*), following the construction of Horns Rev wind farm 'there was statistically significant reduction in the number of bird encounters within the wind farm and in the strip of water 2km around the outside of the wind farm. At distances between 2km and 4km from the outer turbines there was no detectable difference between encounter rates pre- and post-construction.' Guillemots *(Uria aalge) and* razorbills (*Alca torda*) also showed an increased avoidance of the wind farm area.

Conversely, the species little gull (*Larus minutus*) showed a shift from avoidance to preference for the wind farm area following construction.

Source: Petersen et al (2006)

multiple wind farms are developed in proximity to each other ('cumulative effects'), such as is taking place in the Belgian, German or UK territorial waters or Exclusive Economic Zones (EEZs) of the North Sea.

In terms of successfully mitigating effects on birds from renewable projects, there are few technological solutions available and careful siting represents one of the most effective ways of managing this aspect.

German mitigation response to barrier effects of marine wind farms

Concerns in respect of potential effects from the offshore wind turbines in the German Exclusive Economic Development Zones were of sufficient magnitude that conditions were inserted in the project permissions allowing the regulatory authorities to stop production from the wind farm during large bird migration events. This measure was adopted to reduce the potential for large bird collision fatalities and to reduce the barrier posed by the moving blades. However, the practical application of such a measure remains to be confirmed.

9.2.5 Landscape and visual change

Visual impacts from new developments upon existing landscapes and seascapes can form a significant barrier to the approval and deployment of marine renewable generation technologies. By its very nature, the marine environment often presents very open vistas with an absence of screening terrain and structures. Moreover, during periods of darkness, few background sources of light may exist to provide a backdrop for point-source emissions of light from generation assets (e.g. anti-collision lighting).

The issue of landscape and visual impacts is a very subjective one and depends upon the personal preferences of the viewer. However, opponents of the spread of renewable energy (with particular reference to wind turbines) have been very successful during recent years in harnessing public opinion and gaining media coverage of their objections. In contrast, advocates of renewable technologies in the landscape have been largely ignored.

Landscape and visual impacts from offshore renewable projects are most likely to prove a barrier to offshore wind developments, as the tall turbine structures will be very visible against the skyline. Safety lighting requirements, including anti-collision aviation lighting, will also serve to make these structures more visible. Wave and tidal devices, with their smaller profiles, will be less visible and therefore less likely to provoke an adverse reaction; however, shoreline wave energy devices may also be highly visible. Landscape and visual concerns should provide no barrier to fully submerged devices.

Consideration of impacts of this nature is also most likely to be confined to operational installations as construction activities are normally viewed as being temporary and not, therefore, significant in the longer term.

9.2.6 Marine processes

The removal of energy from the water column by renewable energy generators may also produce impacts upon the marine environment which may prove a barrier to development. Depletion of energy from the water column could lead to localized changes in water movement in the area immediately surrounding generation assets. This, in turn, may lead to alterations to sediment scour and deposition patterns and possible alterations to turbulence and stratification within the water column. Changes of this nature can have an indirect impact upon ecological receptors through habitat modification and also upon the human environment through changes to patterns of sedimentation.

Where a renewable generation site is located in a near-shore environment, alteration to patterns of sedimentation may affect beach replenishment and erosion mechanisms, with implications for existing coastal defences and coastline management. Moreover, if the site is located in proximity to navigation channels, sedimentation changes could affect requirements for dredging and/or present a safety hazard.

Recent research

A recent study by Boehlert and Gill (2010) highlights the following in terms of removal of energy from the water column:

> In the far field, energy reduction could lead to changes in currents and subsequent alterations in sediment transport. Although few studies have been undertaken, surveys at an installed wind farm in the North Sea that used monopole foundations with scour protection showed secondary scouring (Rees et al., 2006). Further, a modelling study based on wind farm data highlighted far-field deposition downstream of the wind turbine foundations (Besio and Losada, 2008).

In respect of tidal energy, removal of sufficient energy could theoretically result in localized modification of the tidal range, affecting intertidal ecosystems. Research with regard to potential effects on the water column and seabed morphology is on-going and it is likely that more work will be required before any conclusions can be drawn.

9.2.7 Historic and cultural assets

The marine environment has the potential to present a significant resource in terms of historic and culturally significant artefacts. In particular, shipwrecks and aircraft crash sites can contain a unique assemblage of artefacts from a specific time or event in history. In addition to man-made artefacts such as ships and aeroplanes, prehistoric landscapes have also been drowned by rising sea levels and are preserved as buried paleo-landscapes. The North Sea represents one such drowned landscape and prehistoric animal remains are regularly brought up in fishing nets, as are occasional prehistoric human artefacts such as stone tools and worked bone. The inherent difficulty which is often encountered in accessing such sites can lead to a high degree of preservation of individual objects, which can frequently be enhanced by burial in marine sediments.

Many shipwrecks and aircraft remains also have a cultural significance above the value of the artefacts preserved within them in that many such sites may contain human remains. Moreover, it should be noted that military wrecks remain under the jurisdiction, and often under the protection, of the government that lost the ship or aircraft or that government's successor. Many military wrecks are protected by virtue of their being classed as war graves, with disturbance of such sites strictly prohibited.

All types of marine generation asset placed offshore have the potential to affect submerged or buried historic and cultural assets. This can be a particular consideration when placing sub-sea cabling, where cable routings can cover many kilometres of seabed and are regularly amended to avoid seabed features and remains.

9.2.8 Change of use conflicts

Change of use conflicts can present another notable non-technical barrier to the development of marine renewable generation. Marine areas are already heavily used for many business and recreational activities, such as fishing, oil and gas activities, sand and gravel extraction, aquaculture, recreational diving and sailing, commercial shipping, and military exercises. The placing of renewable energy generators in a marine area often requires the cessation or relocation of previously existing activities and this can lead to confrontation and dispute depending upon the ability and willingness of the other users to move.

Where change of use conflicts arise as a result of competing commercial interests, there is an increased potential for an economic remedy to be sought. This could often entail the provision of either full financial compensation for affected parties to cover the costs associated with relocation, or cessation of the activity, or partial compensation in the form of a subsidy to cover a portion of lost earnings.

Competing uses are addressed in more detail in section 9.8.

9.2.9 Marine safety and communication systems

Interference with marine communication systems is a phenomenon largely associated with offshore wind farms and arises as a result of interference on radar and radio signals from the rotating blades of the wind turbines. While this localized disturbance can create safety concerns, technological solutions exist to mitigate the problems and these can extend to placing additional repeater stations around wind farm sites and using software upgrades to remove interference patterns.

All marine technologies pose some level of navigational hazard to vessels. This is an unavoidable side effect of placing structures in the marine environment and can be largely mitigated during normal activities through adequate lighting, foghorns and the placing and enforcement of exclusion zones around generators. Nevertheless, a requirement will exist for vessels to access sites regularly for maintenance purposes, which will increase the potential for a collision to occur. In addition, during emergency situations (such as a vessel losing power and beginning to drift), fixed and floating structures will also present a hazard. Collision risks will therefore never be eliminated; but as for any other offshore structure, mitigation measures are considered and can be implemented to reduce risks, both in terms of likelihood and severity, to acceptable levels.

Marine safety is addressed in more detail in section 9.4.

9.2.10 Protected areas

Over the last few decades, our growing understanding of the marine environment, and our effects upon it, has led to the designation of an increasing number of Marine Protected Areas (MPAs). As with protected areas on land, these have been designated for a number of reasons, including importance for wildlife, rare habitats and landscape/seascape preservation. As a result, the protection of an area does not automatically mean that it is unsuitable for renewable energy projects; however, it may limit the technologies which would be considered to be appropriate or the scales of the development. In an area designated for its landscape/seascape value, submerged wave or tidal generators would provide a good example of technologies which may be appropriate, while offshore wind turbines would be likely to generate significant opposition.

In respect of protected areas and marine renewable generation, careful and considerate site selection of projects will provide the most effective method of ensuring that significant opposition is avoided. Notwithstanding this, it is likely that any proposal for development within a protected area will incur opposition from some sectors of the community and non-governmental organizations (NGOs). Environmental protection agencies are in their majority supportive of global renewable energy developments in recognition of their key role in combating climate change, but remain vigilant at a local project-specific level (see the following case study).

Extracted view from Scottish Natural Heritage (SNH) Policy Statement No 01/02

Scottish Natural Heritage (SNH) supports the development of renewable energy as an integral part of the government's climate change programme.

Renewable energy developments have the potential to affect valued elements of the natural heritage. The frameworks of policy, planning and support need to help to get the right technologies, and the right kinds of developments, in the right places so as to minimize impacts upon the natural heritage.

Source: http://www.snh.gov.uk/docs/A327600.pdf (Policy Statement No. 01/02)

9.2.11 Environmental impact assessment criteria

The environmental impact assessment (EIA) is the pivotal study required to obtain a permit for development and is intended to assess the environmental impacts of the proposed project and to form a basis for a decision on whether the project should be allowed to proceed.

Several criteria for assessing a project's risk to the environment have been called into question. The first is the precautionary principle which states that if an action or policy has a suspected risk of causing harm to the public or the environment, in the absence of scientific consensus that the action is harmful, the burden of proof that it is not harmful falls on those taking action. In practice, in the absence of data proving otherwise, projects must accommodate a reasonable worst possible scenario.

The second issue is that under a number of EIA regimes, such as the EU, the risk of a proposed project is judged solely on the activities of the project and does not include important externalities, such as the consequences of not pursuing the project, which in the case of not deploying renewable energy would include the social and environmental impacts of fossil fuel extraction, processing and combustion.

In combination, the two issues present a scenario in which the worst case impact of renewable energy deployment is contrasted against an unrealistic best possible case of not proceeding with projects, which is no environmental or social impact for fossil fuel consumption.

9.3 Mitigation and removal of environmental barriers

9.3.1 Marine spatial planning best practice

At national and international level, one mitigation measure in order to ensure the minimization of environmental barriers remains the application and implementation of best marine spatial planning practices.

The role of strategic environmental assessments (SEAs), as defined in the context of European Directive 2001/42 EC, is 'to provide for a high level of protection of the environment and to contribute to the integration of environmental considerations into the preparation and adoption of plans and programmes with a view to promoting sustainable development'. Conducting SEA in advance of planned development or programmes (such as large-scale renewable energy developments) allows the early identification of environmental barriers at national level and the preparation of associated mitigation measures. SEAs also bring important benefits by facilitating future EIAs. The organization and implementation of SEAs dedicated to the offshore renewable industry would allow the identification of areas most suited for project development, as well as provide an analysis of any knowledge gaps which can then be addressed well ahead of further project development through dedicated marine survey campaigns.

The knowledge gaps associated with the interaction of technologies and projects within the natural environmental are the focus of a large number of publicly funded research and development projects and initiatives at national and international levels, as described in the previous section.

Examples of some environmental-specific research and development (R&D) activities, including publication of best practices and description of mitigation measures, can be further consulted within:

- Supergen Marine (2010), COWRIE (2006) and others in the UK;
- RAVE (2010) research programme in Germany.

Collecting data from existing sites, sharing environmental information and collaborating between developers active in neighbouring areas should be actively encouraged by authorities, if not already occurring as a consequence of competitive or confidentiality pressures.

9.3.2 Environmental impact assessment

At project level, the identification, mitigation and removal of environmental barriers is described within the EIA.

An EIA is designed to identify and assess the impacts associated with the development of a project, including:

- construction and commissioning;
- operation; and
- decommissioning.

The EIA process follows a series of well-defined steps which can be broadly summarized as follows:

- scoping;
- environmental baseline study;
- impact assessment and mitigation; and
- preparation of an environmental statement (ES).

These steps are described in further detail below.

Scoping

The initial phase of an EIA, called scoping, involves the determination of those environmental issues which are of potential significance and should be addressed in the EIA, and those which are not of significance and should be scoped out. In addition, the scoping stage should also be used to identify how the EIA will be undertaken both from an overall perspective and from the perspective of assessing individual topic areas. The scoping exercise should also define what information and data should be collected. Scoping a project properly ensures that the EIA focuses on significant and relevant impacts and limits the overall scale of the work to be undertaken, enabling an effective and efficient assessment.

As part of the scoping process, a scoping report is produced, which is used by the project developer to inform project stakeholders and consultees of the details of the project and its potential impacts. Used in this capacity, the scoping report forms the basis for consultation with regulators and other interested bodies prior to the commencement of the main impact assessment process.

Stakeholder consultation

Consultation with key stakeholders (i.e. statutory and non-statutory consultees and the public) is a key part of the EIA process, both for information gathering and for disseminating information regarding the proposals and the EIA process. Consultation should be an on-going process undertaken throughout the EIA and the results of consultation should be incorporated within the assessment process in an iterative manner.

Effective consultation with stakeholders provides one of the best opportunities to identify and address non-technical barriers at an early stage and to minimize the potential for encountering significant non-technical issues at a more advanced point in a project's development.

Environmental baseline study

Any marine renewable energy project EIA will require a thorough investigation of the existing environmental conditions at the site and in the area of potential influence: the environmental baseline. This will include an investigation of any terrestrial aspects and may include undertaking site surveys and discussions with statutory consultees and key stakeholders to agree on assessment scope and methodologies.

The findings from the baseline investigation should be fed into the relevant specialist topics of the EIA and form the basis of the impact assessment and final ES submission. Environmental baseline components that we would normally expect to be described would include, but are not be limited to:

- climate and topography;
- marine environment (including sub-tidal, inter-tidal/coastal as applicable);
- noise (marine and surface);
- environmentally sensitive areas, including existing and proposed areas;
- occurrence of rare and/or protected species;
- archaeology and cultural heritage;
- terrestrial environment;
- fisheries;
- other marine users (recreational, shipping, etc.);
- socio-economics;
- seascape/landscape; and
- water and air quality.

Impact assessment and mitigation

Impact assessment and identification of mitigation measures should be undertaken by environmental specialists using established EIA techniques, and in line with applicable legislation and guidance (taking account of any relevant international lender requirements such as the Equator Principles). All assessments should consider normal, abnormal and potential emergency conditions for all phases of the project life cycle and an assessment of cumulative impacts should also be included.

The assessment of each topic should follow the key steps outlined below:

- identifying and describing potential impacts during construction, operation and decommissioning;

- identifying the sensitivity of baseline receptors to predicted impacts;
- describing the nature, extent and magnitude of predicted impacts;
- evaluating the significance of predicted impacts (prior to the adoption of mitigation measures);
- identifying appropriate mitigation measures to avoid/minimize and/or reduce potentially significant impacts;
- identifying residual impacts and evaluating the significance of these impacts post-adoption of mitigation measures; and
- recommending monitoring strategies where required.[1]

Preparation of the environmental statement

The output from an EIA is an environmental statement (ES). This document presents the findings of the EIA and should be accompanied by a smaller non-technical summary (NTS) providing a simple overview of the findings of the assessment.

A typical environmental statement contents list would include the following:

- non-technical summary;
- introduction;
- legislative framework;
- project description and alternatives;
- environmental baseline description;
- environmental and social impact assessment;
- mitigation measures;
- conclusions and recommendations.

9.4 Health-and-safety barriers

While health-and-safety compliance requirements are subject to country-specific legislation, the overarching requirements of safety at sea and safety at work are universally applicable. In this section, we specifically focus on safety issues associated with the construction and operation of offshore renewable energy projects in the marine environment. Other safety issues and concerns are linked to onshore activities (e.g. at ports), such as manufacturing, transport, construction or operational activities, but these will not differ in essence from other onshore activities and are therefore not specifically covered in this section.

The main safety issues specifically faced by offshore renewable energy technologies and projects can be summarized as:

- construction and operation activities (vessel suitability, vessel stability, lifting operations);
- emergency planning and response;

- litigation and reputational risks;
- collisions risks (vessels, aircraft) and signal interference;
- access and personnel transfer (from boats and helicopters);
- weather (wind speeds, wave heights and tide conditions and envelopes for safe operation) and distances on project planning (passage planning, shelter, weather windows for construction and maintenance);
- divers and sub-sea work;
- electrical safety;
- management of safety interfaces between stakeholders (owners, contractors);
- availability of appropriate standards and guidance (as emerging technologies and existing standards applicable to other industries such as the oil and gas industry may not be directly applicable to or suitable for offshore renewable energy technologies);
- skilled personnel (general lack of experience and knowledge from the industry, and shortages of skilled personnel and crew);
- training of personnel for routine and emergency situations.

A strong record of safety compliance, associated with evidence of a good safety culture and certification to health and safety standards (such as OHSAS 18001), is usually high on the internal agenda of project developers and utilities. External contractors are required to demonstrate these qualities in order to qualify for any offshore work or tender. While, on the one hand, these requirements are essential in order to ensure a very good track record of the industry as a whole, they can, on the other, act as a barrier for companies to enter this market, particularly companies not already operating in other offshore industrial sectors such as the oil and gas or the maritime industry.

The United Nations Convention on the Law of the Seas outlines that within its Exclusive Economic Zone, a coastal state has the exclusive right to construct, authorize and regulate construction and operation of installations and structures (Article 60). A number of international conventions are in place and are applicable to offshore renewable energy projects developed within the EEZ signatory countries, a list of which is published by the International Maritime Organization (IMO, 2010). Some of the main international conventions related to safety and applicable to offshore wind, wave and tidal projects include (Renewable UK, 2010):

- Solas 1974 (Safety of Life at Sea);
- STCW 1995 (International Convention on Standards of Training, Certification and Watchkeeping for Seafarers); and
- COLREG 1972 (Convention on the International Regulations for Preventing Collisions at Sea).

National and transnational (e.g. EU directives) regulations and compliance requirements will also apply and introduce particular permit conditions requirements, among others, and the assurance of navigational safety, usually including official communication of the exact location offshore structures and specific marking requirements for site locations and devices.

Permit applications procedures and requirements often require the developer to provide evidence of compliance to health-and-safety legislation, assessments of risks and availability of mitigation measures or plans. For example, the German offshore wind permitting process requires the submission of a state-of-the-art collision risk analysis and consequence of potential pollutant spill, and demonstration of compliance with three standards specifically developed by the Bundesamt für Seeschifffahrt und Hydrographie (BSH, 2010).

While the probability of a collision between a vessel and an offshore renewable energy structure occurring is low (given the siting of offshore parks and mitigations measures in place), the impact of a potential collision might be large if it results in the loss of vessel, life or permanent damage or losses to generation equipment.

9.5 Mitigation and removal of health-and-safety barriers

9.5.1 Project-level mitigation measures

Table 9.1 summarizes the main health-and-safety issues and provides examples of some typical mitigation measures which can be developed in order to reduce or remove the safety issues and risks identified in the previous section.

At project level, the existence of an effective safety management system needs to be demonstrated. Such a system must describe how risks are continuously identified, assessed and either managed, eliminated or mitigated, usually to 'as low as reasonably practicable' levels. Well-established safety techniques and methodologies (such as hazard identification – HAZID) are in place to support the safety assessment process. The safety management system will be described in a project-specific safety document which should also demonstrate and describe the plans that are in place to manage major accidents, and highlight responsibilities, including names and contact details.

In many countries, detailed collision risk assessment studies have to be performed as a preliminary permitting requirement for offshore structures. For example, a Marine Navigational Safety Risk Assessment is to be produced as part of the EIA in the UK. In Germany, the permitting process requires the submission of a state-of-the-art collision risk analysis and consequence of potential pollutant spill (BSH, 2007).

Practice shows that safety zones have been introduced to mitigate collision risks during construction. Establishment of a safety zone during operation has been discussed for some technologies (e.g. floating devices). Today various developers

Table 9.1 Main health-and-safety issues and typical mitigation measures

Issues	Mitigation Measures
Construction and operation activities	Selection of appropriate vessels, contractors, personnel Planning Monitoring and forecasting of weather conditions Evidence of good safety record and culture in contractors OHSAS 18001 accreditation of contractors
Emergency planning and response	Emergency plan and procedures in place within health and safety management system Shelter areas
Collision risk and navigational safety	Collision risk study Design of structures Exclusion zones and siting Signals and markings
Access and personnel transfer	Specialized vessels and access procedures Staff training and qualification Personal protective equipment (PPE) Personnel tracking
Weather	Monitoring and forecasting of weather conditions (wind speeds, wave heights, period and direction, tidal range and flows, temperature) Understanding of acceptable envelopes for safe operation or construction
Divers and sub-sea work	Competency and training Suitable and specialist vessel and equipment Remote operating vehicles (ROVs)
Electrical installations	Personnel competency and training Signage Restricted areas
Safety interfaces	Interface matrices and structure clearly define respective roles and responsibilities between owners, operators and contractors
Appropriate standards and guidance Legal requirements	Development of specific guidance and standards by the industry, regulators, certification bodies and international organizations National and international dissemination of best practices National and international harmonization
Availability of skilled personnel	Development of training courses and certification Campaigns to promote career prospects within education systems Programmes to support and facilitate transfer of skills from other (declining) industries

implement vessel traffic services based on automatic identification systems (AIS) on their offshore installations, in addition to compulsory sea markings in order to increase navigational safety (Biehl and Lehmann, 2006).

Various institutions have undertaken assessments of collision risks for ships with offshore structures and offshore wind turbines, in particular (GL, 2002; Den Boon et al, 2004). More specifically, requirements for the collision-friendly design of offshore wind turbines and their foundation structures have been defined. Computer models and tests with scaled models have been undertaken to analyse the collision risks and impacts to offshore wind turbines. Such tests indicate that three major parameters determine the impact of the collision (i.e. spacing of wind turbines, foundation type and whether the vessel is powered or drifting). If turbine spacing allows the ship to pass safely through the wind farm, there will be no collision and no material impact upon wind farm operation. Collision behaviour of the foundation varies depending upon the type of foundation (i.e. monopiles are expected to fall away from a colliding ship, whereas there are scenarios where tripods are expected to drop the wind turbine onto the colliding ship) (Biehl and Lehmann, 2006).

Another example of a mitigation measure in place in order to reduce the collision risk caused by a drifting vessel can be found in permit conditions attached to German offshore wind farms. These include the requirement for offshore wind farm owners to cooperate logistically and financially in the availability of tug boats, which can be mobilized at short notice in order to assist vessels adrift. This permit condition becomes enforceable in the German North Sea sector once a given cumulative capacity has been installed.

Experience regarding design requirements in offshore construction has been gained in the oil and gas as well as the infrastructure sector. Consequently, a range of design codes has been introduced with particular regard to offshore construction; these include international standards such as the ISO and IEC, as well as national standards such as the Norwegian Petroleum Directorate. These standards define design requirements, including factors such as the minimum load to be taken by the structure in the case of collision. Further standards and assessments have been discussed in detail in the publications from the International Ship and Offshore Structures Congress highlighting the different approaches to the estimation of likelihood of incidents, the assessment of risks and the mechanics of collisions (ISSC, Committee V.1, Collision and Grounding, 2006).

9.5.2 Other initiatives and mitigation measures outside project level (industry, national and international)

A number of initiatives and other measures can be taken by industry, governments and international institutions in order to contribute to the continual improvement of the offshore renewable industry's health-and-safety record and performance. The following case study highlights the initiatives which are taking place in the UK.

Case study Health-and-safety improvement initiatives in the UK offshore renewable energy industry

In the UK, renewable energy industry trade associations, industry and regulator (the Health and Safety Executive) are working closely together and have launched a number of initiatives to contribute to the continual improvement of the industry's health-and-safety record.

Some of these initiatives include:

- development and dissemination of industry-specific guidelines, such as the recently revised *Guidelines for Health and Safety in the Wind Energy Industry*, the *Guidelines for Health and Safety in the Marine Energy Industry* or the *Guidelines for the Selection and Operation of Jack-Ups in the Marine Renewable Energy Industry* (Renewable UK, 2010);
- gathering of information on safety accidents, incidents and near misses by Renewable UK and, subject to confidentiality provisions, issue of quarterly reports (in some significant cases, issue of 'Safety Alert' to all trade association members to ensure maximum and immediate dissemination of lessons learned and recommendations);
- development of a safety accreditation scheme in association with training providers;
- industry committee on health and safety;
- contribution to the development of international standards;
- development by the Health and Safety Executive (HSE) of a dedicated Emerging Energy Technologies Programme in order to 'bring together the ongoing work and any future projects to produce (and to share with our key regulatory partners) a coherent organisational strategy for HSE's regulation of the emerging energy technologies; and guidance to enable HSE's divisions to plan and deliver against this strategy' (HSE, 2010);
- organization of dedicated health-and-safety conferences and open dissemination of presented material.

The direct availability and applicability of a large number of offshore standards and guidance documents have already been mentioned. The main organizations publishing such standards include:

- the International Maritime Organization (IMO);
- certification bodies and standards organizations such as the IEC, Germanischer Lloyds (GL), Det Norske Veritas (DNV);
- national regulators; and
- industry trade associations.

These organizations have also all recognized the need for industry-specific guidance and standards to be developed, and many of them have already published such material.

Technical innovations (dedicated vessels, new procedures) are being developed in order to improve the accessibility to offshore installations and the safety of personal transfer in a larger envelope of metocean (wind, wave, tidal) conditions. The same applies for sub-sea work and interventions, where remote operating vehicles (ROVs) can, when applicable, be used instead of divers.

9.6 Regulatory and permitting barriers

Offshore renewable energy is on a steep commercialization curve and is vulnerable to a raft of regulatory and permitting hurdles that impede its prospects for success or the extent to which the technology can diffuse in the market and have a meaningful impact upon reducing fossil fuel dependency.

From experience, projects involving the following scenarios are subject to a high degree of policy and regulatory impact and often encounter very challenging or insurmountable regulatory and permitting hurdles:

- environmental technologies which rely on policy and regulation to create market demand;
- major infrastructure projects which are subject to strict and complex planning requirements;
- nationally driven projects which are implemented on a local level, triggering conflicting local and national interests;
- attempts to significantly alter the commercial arrangements of utility sectors, which are highly regulated, often operate under regional monopolies and are constrained by a set of strict regulatory remits; and
- any initiative which brings into play the management and allocation of natural resources, especially where multiple government jurisdictions and interests are involved.

Any of these challenges can cripple the progress of innovation, and offshore renewable energy is subject to all of these challenging scenarios.

In addition to these sector-related regulatory challenges, offshore renewable energy encounters a number of innovation-related regulatory challenges, including:

- an inability of an innovation to meet prescriptive standards or regulations that were developed around conventional approaches;
- gaps or uncertainty in legal issues or policy;
- a lack of a regulatory authority or consistent policy;

- inadequate fiscal instruments;
- competitive hurdles such as trade barriers or restrictive standards; and
- lack of government expertise and resources to address new and complex issues.

These barriers arise in different ways, in different regimes, but tend to be most relevant in the context of:

- the regulatory framework;
- securing seabed rights and permits;
- promoting innovation; and
- securing grid connection and power distribution.

This regulatory track parallels the key project development mileposts of:

- assessing project viability;
- securing the rights to develop a site;
- designing and constructing the infrastructure; and
- delivering power to the grid.

Regulatory barriers can undermine any of these key milestones and endanger project success.

9.6.1 Regulatory framework

The lack of a robust and reliable regulatory regime is one of the primary reasons why projects are determined not to be financially viable and do not attract investment. In emerging economies, projects which are otherwise promising may remain un-bankable due to a lack of a strong regulatory framework, currency risks and a lack of supply chain capacity. In developed countries, government programmes for renewable energy can also be tenuous and vulnerable to shifts in party leadership, the economy or public sentiment. In the US, production tax credits played an important role in supporting wind energy during the late 1990s. However, they were subsequently allowed to expire in 1999, 2002 and 2003 resulting in a 'collapse' in wind energy development (WRI 2009). Even in countries with strong regulatory drivers, such as EU member states, policies can rapidly change, as in the case of Spain, which recently announced reductions to wind energy feed-in-tariffs.

The degree to which a country's policy is underpinned by the political and economic landscape is also relevant. For example, in countries such as Norway, a lack of legal framework for permitting,[2] coupled with a large untapped potential for onshore wind and hydropower available at a lower cost, has so far acted as a barrier to the deployment of offshore renewable energy technologies.

In the UK, government is under pressure from numerous European Commission (EC) commitments, and its own internal commitments to reduce greenhouse gas (GHG) emissions and increase renewable energy production. However, unlike other nations, options for onshore renewable energy are limited due to the exceptional difficulty in obtaining planning permissions and achieving grid connections. In the case of the UK, offshore developments offer ways around these barriers and in theory capitalize on strengths in the dwindling offshore oil sector.

Norway's policy landscape

In Norway, the push for offshore development has been tempered by the vast areas of land available for less expensive onshore renewable energy. While onshore development has been granted support by relevant stakeholders, applications for offshore systems have been refused on environmental grounds. The government has taken a more conservative approach relative to other nations in order to explore options, having drafted legislation on offshore Renewable Energy Production, Proposition No 17 (2007–2009), to assess areas that may be suitable for offshore deployment.

At the same time, the country has taken an innovative and more aggressive approach in developing the world's first large-scale floating turbine, which can be moored at a great distance offshore with no visual impact and potentially less disturbance to marine ecosystems (CCA, 2010).

A favourable top-down policy is important in laying the groundwork for offshore development. However, it will be of limited value in accelerating deployment regulations if policy and processes have failed to penetrate laterally through central government agencies and vertically through regional and local governments. Central government agencies tend to operate in silos under sets of narrowly defined policy remits, and local and regional government policy is often aligned with local and regional priorities, rather than national policies. Without a central framework in place to drive national renewable energy policy across multiple agencies on a national, regional and local level, project development prospects can flounder amid uncertainty and delay.

The metric used to determine a strong regulatory framework is somewhat subjective but important. The attractiveness of a regulatory framework is not assessed by the commercial sector in isolation, but in relation to other project risks and international investment opportunities within and without the offshore renewable sector. In a sector where demand exceeds supply and commercial players operate internationally, developers and investors have the option of choosing the most attractive location for investment. Government mechanisms, such as tariffs and permitting programmes, which are, in principle, adequate, may not be effective in attracting investment when combined with other risk factors, such as technical or supply chain constraints, and compared to less risky opportunities elsewhere.

National regimes may also fail to attract investment if the regulatory regime combined with other local factors does not encourage development on a scale significant enough to justify the development of supply chains and necessary grid networks that are required to support project development. Because of the significant local supply chain development required, a pipeline of promising projects is necessary to justify full-scale offshore renewable energy investment.

Finally, removing political risk is a key factor. Investors and developer need to have confidence that the regulatory regimes will not be unduly changed.

9.6.2 Financial support schemes

In general, the level of support for offshore renewables tends to be in line with onshore wind and bioenergy support and less than that for photovoltaic (PV). In fact, few jurisdictions make any special allowance for the higher costs of these immature technologies, though differentiation is increasingly being applied for offshore wind as opposed to onshore wind.

The support schemes primarily include feed-in tariffs and tradable green or renewable energy certificates, but may also include tax incentives and, potentially, instruments such as soft loans, grants and loan guarantees.

The effectiveness of these support schemes cannot be judged purely on a numerical basis against levelized costs, but must take into account the stage of technology development, the financial markets and the objectives of balance sheet investors.

9.6.3 Allocation of seabed rights

The allocation of seabed rights is an important milestone in the development of offshore renewable energy.

Territorial waters, Exclusive Economic Zones and the continental shelf

Under the United Nations Convention on the Law of the Sea (UNCLOS): 'A coastal State has sovereign rights for the purpose of exploring and exploiting the natural resources of its continental shelf, and the exclusive right to erect structures or installations for these purposes' (DTI 2002). Separately, the convention gives a coastal state the right to establish a 200 nautical mile (nm) Exclusive Economic Zone (EEZ) around its territory with which, among other things, it can exercise sovereign rights in the production of energy from the water, currents and wind. Territorial waters extend 12nm out to sea.

Each nation will have its own system under which it governs these offshore waters. Many nations such as Poland, Denmark and Ireland treat their territorial and EEZ waters the same for permitting purposes. Others such as the US have a system where waters out to 3nm are state controlled and beyond that are considered federal waters managed by the Department of the Interior. In the UK, the Crown Estate has authority for the territorial waters seabed and under the Energy Act 2004 also has the powers to license energy developments within the EEZ.

Jurisdictional barriers to seabed rights

As the offshore energy sector matures, near-shore sites will become scarcer and developers will increasingly be forced to look further offshore. This has already been the case in the UK, Norway and Germany, all of which have identified potential development areas which run along their EEZ borders. Inevitably, at some point in the future, an offshore energy development will span national jurisdictions. At this time we are not aware of any arrangements in place in Europe to deal with such an eventuality, although the recently commissioned Robin Rigg offshore wind farm built in Scottish territorial waters actually exported its electricity to a substation in England, thus requiring consent from English and Scottish authorities.

Disputed EEZs are also an impending issue. EEZs are a fairly recent introduction into international law (circa 1970s) and as such there are still a number of disputed areas of water around the world. Of the countries being studied in this book, Japan is one of the most affected by this issue, as are many other nations in the crowded South China Sea. Japan has a number of areas disputed between itself and South Korea, Russia, Taiwan and China. Historically, these disputes have curtailed Japan's planned offshore oil and gas activities and could well affect future offshore energy developments. Around half of the countries considered in this book have an EEZ dispute, of varying size, with a neighbouring country.

Licensing barriers to seabed rights

The process of granting licences varies by country; but there are two broad approaches to granting licences for investigation or development. The first approach is the 'developer' (also referred to as the 'first come first served') approach and the second is the tendered approach. Under the developer-led approach, site developers apply for a series of licences that are eligible for approval on a first-come-first-served basis. Under the tendered approach, the government requests a proposal to develop a designated site and may undertake preliminary works to establish site conditions and address important site risks.

The site development approach can potentially give rise to several barriers, especially in countries with higher risk sites. The primary hurdle for the developer is the cost and risk associated with investigating potential sites and undergoing the permitting process, especially if the government's policy position on site development and other relevant issues, such as grid capacity, is vague.

The risk for government policy-makers is that the parties who venture down the path to permitting and development may not be the best qualified in technical or financial terms, and in the long run valuable time could be lost if the developer cannot carry the project through to completion.

In practice, what may occur in developer-led sites is a commercial party may de-risk the site and sell the rights to develop the site to a third party. This has raised concerns that the developer-led model may attract parties with a permit focus rather than construction focus (i.e. speculative reasons) (WRI 2009). Technical specialists have noted that studies undertaken in development-led models may need to be repeated because they were not comprehensive enough to address broader site development concerns.

Ireland's experience

Before 15 January 2010, developers could apply to the Sustainable Energy Authority of Ireland for a foreshore licence at a fee of €100,000, which entitled them to investigate a site for up to four years, after which a further €900,000 deposit could be paid as a reservation for development. The authority reserved the right to deny this application and, if so, the next developer in the queue could apply for a foreshore licence and continue the process (SEI, 2001). This led to a secondary market where a small company could buy the foreshore licence, and then having produced a small environmental impact statement, potentially sell on the foreshore lease at a profit.

As of January 2010, authority for foreshore licence applications has been passed to the Department of the Environment, Heritage and Local Government. Now any developer wanting to apply for what is called a foreshore consent must engage in a pre-application consultation, and stricter rules have been established on changing a foreshore consent into what is now called a foreshore licence (a 35-year lease and licence for development).

9.6.4 Planning and permitting barriers

Regardless of whether countries apply a developer-led or a tendered licensing process, the procedure for undertaking environmental impact statements (EISs) and gaining approval for offshore development can often be a complex process, with numerous stakeholder consultees or public bodies engaged in the approval process. In the US and other countries, where no single regulatory body has authority over granting development rights, developers must make separate applications to numerous agencies.

The lack of a consistent permitting approach is one of a number of permitting issues that can stack the odds significantly against project success. Other barriers that get in the way of a streamlined process include:

- a lack of a clear permitting pathway;
- over-reliance on bespoke permitting processes;

Bluewater wind development[3]

The Bluewater offshore wind development project in Deleware, US (currently called NRG), required numerous separate permits, including:

- Delaware State Environmental Review;
- Coastal Zone Act Status Decision;
- Coastal Zone Act Permit;
- Coastal Federal Consistency Certification;
- Subaqueous Lands Permits and Leases;
- Wetlands Permit;
- Clean Water Act Section 401 Water Certification;
- Storm Water Permit; and
- Air Quality Permits.

These permits are issued by a number of bodies, including the US Army Corp, the Environmental Protection Agency (EPA), the Delaware Department of Natural Resources and Environmental Control (DNREC) and the Delaware Division of Water Resources. The Delaware permitting process itself is still in flux as illustrated in July 2010 when the EPA delegated authority for issuing Air Quality Permits to DNREC,[4] a move which is hoped will speed up the permitting process.

- overly detailed design requirements;
- a lack of regulator resources or expertise; and
- environmental impact assessment criteria.

These barriers add significant cost, time and risk of failure to individual projects and can substantially slow the progress of national programmes.

Lack of a clear permitting pathway

While permitting processes and regimes differ extensively between countries, this in itself has not been considered a major obstacle to development, provided that the permitting process is transparent and streamlined (EC, 2005). The failure of governments to lay out and evaluate the efficacy of permitting pathways, identify barriers and overcome legal and procedural sticking points will defer investment and pose the risk of programme failure or issues being resolved in the courts.

Over-reliance on bespoke permitting processes

A substantial pipeline of offshore renewable energy projects is required to encourage supply chain development, generate economies of scale and give meaningful renewable energy production. Permitting regimes, which require developers and regulators to undertake non-material repetition of permitting, act as a barrier to development.

An overly bespoke process also promotes an environment of regulator inconsistency and uncertainty, and undermines the ability to prepare permitting methodologies or site technologies in response to a common set of constraints.

Repetitive permitting regimes can be especially burdensome for technologies in the demonstration phase, which require a number of repeated applications with relatively small design changes.

Requiring detailed designs in the permitting stage

Requiring detailed design concepts in the permitting stage presents two pitfalls. The first is that the approach substantially increases the cost and risk of preparing permit applications, and to reduce these costs, the amount of studies and engineering going into initial design concepts is minimized. The second problem is that the specialists employed to prepare permit applications tend not to be design engineers. Once permissions have been granted the ability to change designs can be severely restricted without repeating the lengthy permitting process. This places serious constraints on the design engineer's ability to innovate and reduce costs or even to meet functional design requirements. The areas of concern raised by design engineers include foundations, the laying of seabed cables and specification of boats, which all require detailed design. They also remarked that specifying a specific piece of equipment in the permitting phase undermines one's ability to attract competitive bids; as a result, prices can skyrocket. This may especially be an issue in developer-led licensing models in which rights for development are sold from party to party.

Lack of regulator capacity

Permitting new and complex technologies in an efficient and consistent manner requires specialized expertise and enough experience to develop a well-informed perspective on key issues such risk. While data on regulator expertise on offshore renewable energy deployment are not readily available, a lack of regulator expertise and consistent approach is a common challenge in the implementation of environmental innovations.

In the case of offshore technology, there is also a perceived need for more data to support regulator decisions, especially environmental data (EC, 2005).

9.6.5 Support for innovation

Innovation in design and construction is crucial to reduce costs and impact upon the environment. Regulation and policy can impede innovation in a number of ways, as outlined below.

Barriers to demonstration

Demonstration projects are critical for technology development and instilling market confidence. Regulatory and permitting burdens can be especially acute in the product demonstration stage in which the resources and time required to navigate regulatory requirements and acquire permits may be too onerous, expensive or time consuming for small business or investors, who are already facing significant project risks with little prospect of near-term revenue. Overly burdensome permitting or compliance requirements can work against a government's interest by encouraging businesses to test (and subsequently develop) their products abroad. Common regulatory barriers to demonstration include:

- disproportionate permitting requirements;
- overly prescriptive regulations; and
- requirement that products be certified to industry standards.

Regulations drawn up for full-scale deployment may not include specific provisions for demonstration projects which present significantly reduced or temporary impacts. In a similar vein, other regulations may lay out requirements that are disproportionate for demonstration-scale projects or even overly prescriptive and, hence, prevent new approaches from being tried in a controlled manner.

Any requirement that technologies be certified to industry standards, even for risk or safety reasons, can be especially onerous or impossible for products in the demonstration phase. Gaining full certification to an industry standard is not only very expensive and time consuming, but may be strategically ill advised or even impossible if a production facility has not been established or the product design has not been finalized. Any changes to the production process (assuming that there is a production process) or design resulting from a product trial can make the certification null and void. Innovations, by their very nature, may also not comply with prescriptive requirements laid out in standards developed for conventional approaches. However, standards can provide useful guidance to new technology developers. They can also drive improvements to project and device quality, as well as international consistency.

Cambridge Institute of Manufacturing Innovation Study

An analysis of 100 UK environmental innovators by the Cambridge Institute of Manufacturing found that the most significant barrier to commercialization was not finance for capital or R&D, as might be expected, but the combined challenge of proving technology performance and gaining certification. While this study may reflect the stage of technology development and may not directly translate to the offshore renewable sector, it highlights the need for careful consideration of these issues.

(CIM, 2006)

Barriers to market entry

Overly prescriptive regulations can block out innovations or give rise to confusion if the innovation meets the intention of the regulation but not the letter of the law because of its innovative approach. The application standards can accelerate technology diffusion by instilling market confidence and streamlining processes; however, it can also act as a double-edged sword if not carefully applied. In the case where regulations cite standards, the standards may inadvertently or intentionally block innovation (and competition). Changing an existing standard or creating a new standard to accommodate a new technology is an arduous or even impossible process, particularly if competitive interests have significant influence in the consultation process.

Market entry barriers for innovation (or even mature technologies) can also stem from:

- Trade policies driven by the desire to develop or promote local green economies – an example of this is local content requirements (LCRs).
- Political *quid pro quo* expectations go hand in hand with LCRs and can arise when government representatives channel taxpayer funds towards a preferred industry or major local infrastructure project. Elected politicians may have the expectation that if their constituents' taxes are used in this way, political useful return must be generated, such as local jobs or business. The 'Buy American' provisions of the stimulus package are one example as long-term energy policy is combined with the immediate aims of domestic job creation and economic growth (WRI, 2009).
- Nationalistic policies: China has raised concerns over its direct exclusion of foreign wind producers from large government project auctions (WRI, 2009).
- Choosing technology winners: government policies that attempt to select the best technological approach and support this technology at the expense of others that are intended to achieve the same objective can be risky and distort natural market forces and undermine opportunities for innovation.
- Utility regulations that discourage new approaches.
- Procurement processes that discourage competition.

Highly regulated utility companies, such as grid distribution operators, may be discouraged from innovating by an inability to justify the venture on the basis of the short payback period specified in their contract (typically five years).

Government procurement strategies can encourage monopolies or oligopolies either purposefully, as in the case of regulated utility companies, or inadvertently, as in the example specified below. Since the inception of offshore wind farm development, the number of developers has reduced significantly. Some of the paring down has been a healthy transition to a more qualified and robust industry; however, governments need to remain vigilant in order to avoid market dominance resulting in higher prices and less innovation.

For example, in the Netherlands a single wind farm developer from Germany was selected to develop a number of the Netherlands' largest offshore wind farms. Because of the scale and timing of the project, it enables a single developer to rapidly achieve a greater economy of scale than its competitors and achieve a level of market domination.

9.6.6 Regulatory and permitting summary

Feed-in tariffs and other support mechanisms rely on stable government policy. When combined with other project risks, including permitting, the reward/risk ratio may be too low to attract development on a wide scale. Countries must not only endeavour to streamline their permitting processes and reduce risks to manageable levels, but must ultimately create a financial proposition that is strong enough to draw investors away from lower risk, higher profit investments in other countries or sectors. Providing adequate tariff support is one means of doing this; but other policy measures that promote efficiency, innovation, competition and effective allocation of costs are also important.

9.7 Mitigation and removal of regulatory and permitting barriers

While each region has its unique regulatory and permitting regime, common trends in approaches are beginning to emerge.

9.7.1 Allocation of seabed rights/tendering

The two broad approaches in granting seabed rights to develop offshore sites are the open system (also termed the first-come-first-served system) and the tender system. Countries such as the Netherlands, Ireland, Sweden and Germany have applied the open system, whereas the UK, Belgium and Denmark have applied the tendering system (EC, 2005).

There are several variations on these approaches that have been undertaken by governments to attract investment, manage early development risk and gain some control over the makeup of the development team. For example, a European jurisdiction has taken the open development approach but is considering offering businesses reimbursement for up-front studies and permit costs if development does not proceed. Another approach taken by the conventional power sector in regions such as the Middle East entails governments undertaking early stage project development as a public–private partnership and then tendering a given percentage of the shares of the special purpose vehicle created to the private sector for final development.

Some advantages and disadvantages of each system are outlined in Table 9.2.

Table 9.2 Advantages and disadvantages of seabed rights allocation methods

Option	Advantages	Disadvantages	Examples
Open	Potentially opens up more possibilities for development Provides a possible path to development in cases where government has not pre-screened potential sites More amendable to shallow and near-shore sites where development risks are smaller	Sites are more likely to be acquired by speculative developers Places emphasis of site development on permitting rather than development Potentially lower cost to government, but government also loses some control of development process and players Delays in Implementation Potential for government to be overwhelmed with permit applications	The Netherlands, Ireland, Sweden, Germany
Tendered	Government has more choice over the quality and capability of developers Government can focus on fewer, more targeted applications When combined with a de-risking of the sites, can provide a more attractive option for construction-oriented developers	Requires more government involvement in site selection and studies Allocation of seabed rights needs to take place in one or more successive 'rounds', which may limit developers' participation	UK, Belgium, Denmark
Open with reimbursement	Removes some of the risk of investment in the open system	Yet to be proven; reimbursement may not be an attractive option to developers	European jurisdiction
Public–private partnership (PPP)	Government plays a role in de-risking site development	More expensive for the government	Middle East – conventional power

More mature offshore programmes, such as those in the UK, have adopted the approach of de-risking project development by conducting strategic environmental assessments (SEAs), followed by a tendering process. This enables offshore development programmes to take a more strategic approach to project development and also select developers who are best positioned to complete the project and manage risks.

9.7.2 Permitting

Previously identified specific issues that increase the risk or burden to the permitting process include:

- lack of a clear and usable permitting pathway;
- lack of a consistent policy amongst central and local agencies;
- the cost and time required to prepare applications and undertake consultation;
- lack of regulator resources and expertise; and
- challenging EIA criteria.

Potential measures to address these issues are outlined in Table 9.3.

Table 9.3 Summary of potential mitigation measures for permitting barriers

Barriers	Potential Mitigation Measures	Examples
Lack of clear and usable permitting pathway	Table-top exercise to identify issues Establish written guidance and procedures Establish single regulator to manage all permit applications and align policies of multiple agencies ('one-stop shopping') Enact supporting regulation Make streamlined regulation compulsory Establish body with authority to identify and resolve permitting barriers Consultation and feedback from industry and trade associations	Table top exercise: UK cabinet office Established procedures: Germany One-stop shopping: Denmark, UK (Scotland), Ireland Supporting regulation: UK Compulsory streamlined regulation: European Commission (EC 2005)

Table 9.3 continued

Barriers	Potential Mitigation Measures	Examples
Lack of consistent policy among central and local agencies	National planning regulation and authority Penetration of national energy and climate change goals across government Bring relevant policy and regulation under single agency	National Planning Authority and Policy: UK inter-regional and country-wide planning and operation EU Integrated Maritime Policy Single agency approach: Denmark, UK's Marine Maritime Organization, Marine Scotland
Burden of developing permit applications and undertaking stakeholder consultations	Pre-approved sites Avoid unnecessary repetition Planning requirements linked to project scale	Pre-approved sites: UK
Lack of regulator expertise and resources	Centralized regulator experts Pre-approved sites Developer funds regulators Capacity-building	Centralized regulator experts: US Contaminated land regime Developer funds regulator: Port of Seattle
Environmental impact assessment (EIA) challenges	Assess project impact relative to whole life-cycle impacts of no action on conventional power Pre-approved sites Collect additional marine data Establish uniform methodologies and terms Flexible conservation options such as preserving other areas EIA requirements linked to project scale	Strategic environmental assessments Flexible conservation options (Germany)

To help address permitting delays, the European Commission has established requirements that member states streamline permitting processes. Unfortunately, they have not placed definitions on this requirement (EC, 2005).

There is a growing trend to the so-called 'one-stop shopping' approach to permitting, in which developers make permit applications to one agency, which has control of the permitting process and is in a position to align multiple interests. Developers were confronted with arduous multi-agency permitting conditions in the Netherlands and Ireland until a one-stop shopping approach was adopted. To date the UK, the Netherlands, Ireland and Denmark have all adopted one-stop shopping approaches. In Germany, developers must still work with multiple agencies, but established procedures are published in a set of three guidebooks.

The following case study highlights how, using the devolved powers at its disposal, Scotland has taken a lead in simplifying the planning application process within Scottish waters.

Scotland's first Marine Bill and creation of Marine Scotland

A 'one-stop shop' consent process is now in place for offshore wind and wave and tidal development within Scotland's territorial waters.

First, new legislation, the Marine (Scotland) Act (2010) was introduced (OQPS, 2010). This act 'provides a framework which will help balance competing demands on Scotland's seas. It introduces a duty to protect and enhance the marine environment and includes measures to help boost economic investment and growth in areas such as marine renewables' (Scottish Government, 2010a). One of its main features is to simplify the licensing system by reducing the number of individual consents.

Second, a marine management organization for Scotland, Marine Scotland, was created in April 2009 (Scottish Government, 2010b). Marine Scotland amalgamates the roles previously played by three organizations and its responsibilities extend to a number of areas including planning, licensing, environment, science, energy, fisheries and compliance.

Another important fixture in permitting regimes is the presence of national planning policies and authorities to provide a national strategic approach and help to break the potential deadlock between local and national interests. The case studies below illustrate how difficult it can be for such national policies to be created and the uncertainties which can be introduced as a result of changes in government policies.

Placing permitting responsibility and marine resources under one national agency is an effective means of aligning offshore policy and priorities across various interests. In the UK, offshore permitting is addressed by the Marine Management Organization, which has the authority to align planning policy across interests such as marine fisheries, shipping and wildlife protection. The EU's integrated maritime policy is intended to accomplish a similar objective.

The sheer financial burden and time required to undertake studies, prepare permit applications and undergo substantial stakeholder consultations, including

Developments in national planning policies

Italy currently has specific planning and permitting procedures in each region. National guidelines for renewable energy are being discussed with the objective of harmonizing the regulatory authorization framework across the country.

In the UK, the Infrastructure Planning Commission had been created in 2008 as an independent body in order to examine applications for nationally significant infrastructure projects, including railways, large wind farms, power stations, reservoirs, harbours, airports and sewage treatment works. However, this organization has recently been discontinued following the change of government in the UK and its remit transferred back to central government.

appeals, can be overwhelming for some developers, particularly for small-scale technology demonstration projects. Governments can reduce a significant portion of this burden and risk by undertaking strategic environmental assessments on potential offshore oil and gas development and wind sites. The establishment of the European Marine Energy Centre in Scotland is also intended to reduce permitting burdens for technology demonstration projects. Developers have indicated that, currently, permitting processes must still be repeated even after minor design changes.

A lack of regulator and stakeholder expertise and resources has the potential to place constraints on permitting processes and also leads to regulator inconsistency on decisions such as risk. One way to address this problem is through the use of regulator experts who are intimately familiar with projects, technologies and important environmental issues. The State of Oregon has recommended a stakeholder education programme to help address the new issues associated with permitting wave power projects. Another more controversial approach in cases where additional regulator resources are required is for site developers to fund additional regulators who can dedicate their full attention to the project. While this raises potential concerns over conflict of interest, the approach has been successfully applied on major development projects, such as extensions to the Port of Seattle.

Environmental impacts and the conclusions drawn by the EIA can have a significant influence on permitting outcomes. Potential measures to mitigate EIA barriers, while maintaining environmental protection, are presented in Table 9.4.

Table 9.4 Summary of potential mitigation measures for environmental impact assessment (EIA) barriers

Barriers	Potential Mitigation Measures	Examples
Uncertainty and the application of the precautionary principle	Better environment data on baseline conditions and environmental impacts Standardized environmental assessment methodologies and definitions	Environmental monitoring programme for Marine Current Turbine's Seagen tidal project in Strangford Lough Special Area of Conservation (Northern Ireland) Implementation of marine surveys in the Pentland Firth to address identified data gaps
Significant or unknown environmental impacts	Develop EIA methodologies that take into account the broader full life-cycle impact of not proceeding with development (climate change, SOx emissions) Develop alternative marine conservation areas elsewhere, if possible	

9.8 Competing use barriers

In this section, the constraints and barriers associated with human activities offshore are presented and discussed. Constraints associated with the natural environment (nature conservation) have already been presented in section 9.2 and will not be repeated here.

Open water spaces (lakes, seas and oceans) may at first sight appear to offer large near-empty zones for the large-scale deployment of offshore marine energy technologies and projects. Although it might be true that many human activities and constraints are removed or significantly reduced offshore compared to onshore, a large number of anthropogenic constraints are faced by project developers. Some of these key constraints include:

- commercial and recreational fisheries;
- commercial and recreational shipping and navigation;
- marine archaeology;

- oil and gas exploitation (platforms and pipelines);
- sub-sea communication and power infrastructure (cables);
- marine aggregate extraction industries;
- mariculture;
- tourism and seascape;
- military and defence practice areas and other use;
- radar for civil or military aviation; and
- dumping grounds (waste, munitions).

The above listed constraints can be further classified as 'hard' or 'soft' (and various intermediate levels) according to the difficulty or ease associated with applying mitigation measures. Hard constraints would, for example, include marine wrecks, pipelines, commercial shipping lanes and other permanent infrastructures (offshore oil and gas, cables), while softer constraints could include seascape or fishing.

Typically, exclusion zones will be drawn around hard constraints, restricting development zones available to project developers. Offshore renewable energy technologies and projects are the most recent, 'new', users of seabed and space; consequently, they are often cautiously or negatively considered by existing users eager to preserve their interests.

The following case study highlights the difficulties associated with defining areas suitable for the deployment of offshore renewable energy.

Case study Competing usages and offshore wind farms siting in the German (North Sea) EEZ

Figures 9.1 and 9.2 provide an example of the difficulties associated with the development and spatial planning of new offshore wind energy renewable projects, as illustrated in the case of offshore wind farm development in the German North Sea EEZ. Large areas are constrained as a result of shipping or military use.

From a complex picture involving many existing developments and activities (Figure 9.1), preferred areas for offshore wind farm development were identified. The onus is on developers to undertake the subsequent studies in order to assess the technical and economic viability of planned offshore wind farm projects. Developments outside the preferred areas highlighted in Figure 9.2 are currently not being considered by Bundesamt für Seeschifffahrt und Hydrographie (BSH), the Federal Maritime and Hydrographic Agency, which decides on the approval of offshore wind farm development projects in the German North Sea and Baltic Sea.

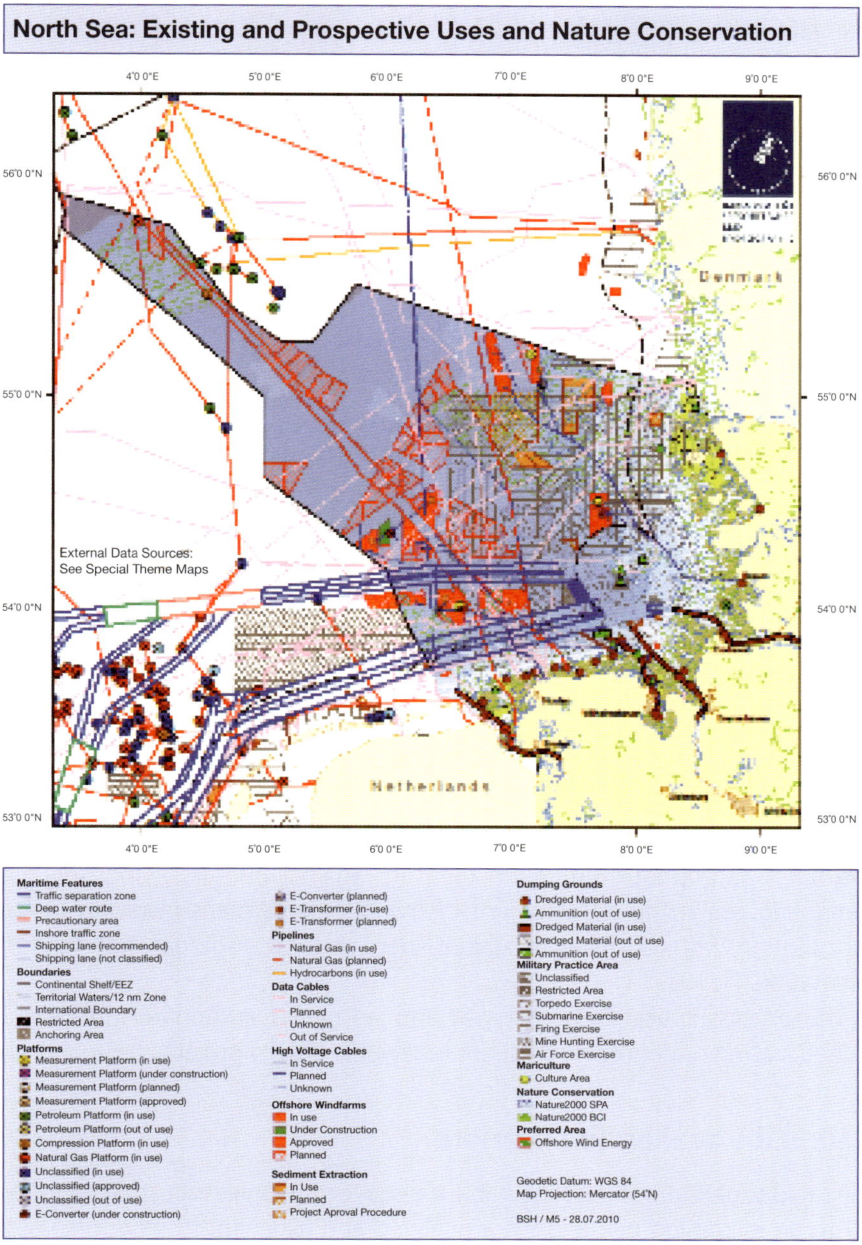

Figure 9.1 Complete uses and nature conservation in the German (North Sea) EEZ

Source: http://www.bsh.de/en/Marine_uses/Industry/CONTIS_maps/NorthSeaCompleteUsesAndNatureConservation.pdf

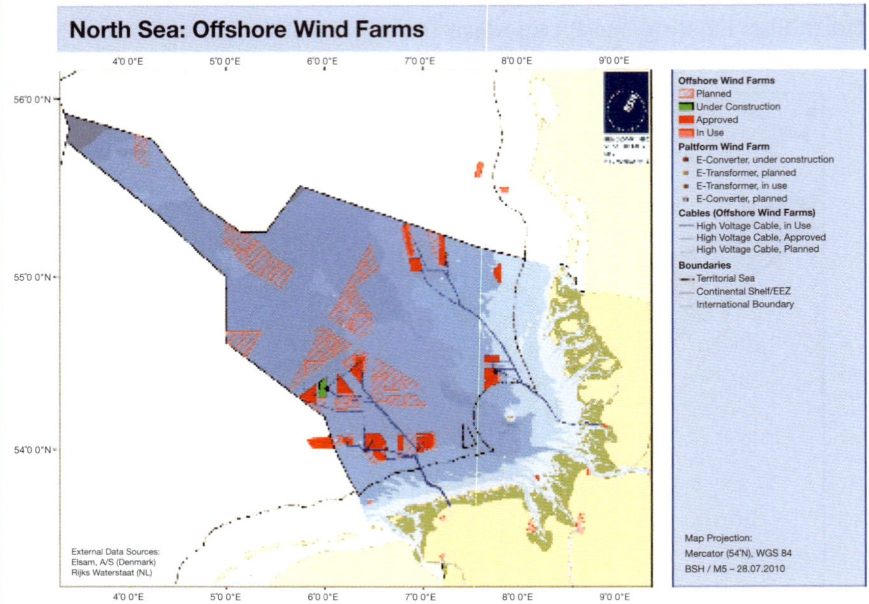

Figure 9.2 Offshore wind farm developments in the German (North Sea) EEZ

Source: http://www.bsh.de/en/Marine_uses/Industry/CONTIS_maps/NorthSeaOffshoreWindfarmsPilotProjects.pdf

9.9 Mitigation and removal of competing use barriers

The most appropriate mitigation measure against barriers associated with competing use is the development of marine spatial plans involving, and in consultation with, all relevant stakeholders at local, regional, national and international levels.

This process is best to be initiated and led by the authorities in charge of seabed leasing or permitting offshore renewable energy projects.

9.10 Skills availability barriers

The rapid expansion of the renewable energy technologies industrial sector has led to an increasing demand for a qualified and experienced workforce at all qualification levels. Skills shortages, particularly in relation to engineering (structural, electrical, mechanical, offshore), construction and project management disciplines (see Figure 9.3) have been identified by industry as posing potential serious barriers to the development of renewable energy technologies, in general, and offshore renewable energy technologies, in particular (BWEA, 2008a).

The following factors and issues can contribute to the skills shortages and act as barriers:

- rapid expansion of the renewable energy sector;
- competition from other sectors of the economy;
- competition from within the renewable energy sector, leading to wage inflation and 'head hunting' recruitment strategies;
- decreasing talent pool as a result of demographic trends, potentially accentuated by a drop in the uptake of science and technology subjects in secondary, higher education or vocational courses;
- lack of awareness of the employment and carrier opportunities in the offshore renewable energy sector;
- difficulties associated with retraining workforce from other sectors of the economy;
- lack of suitable training courses and mismatch between course content and industry requirements.

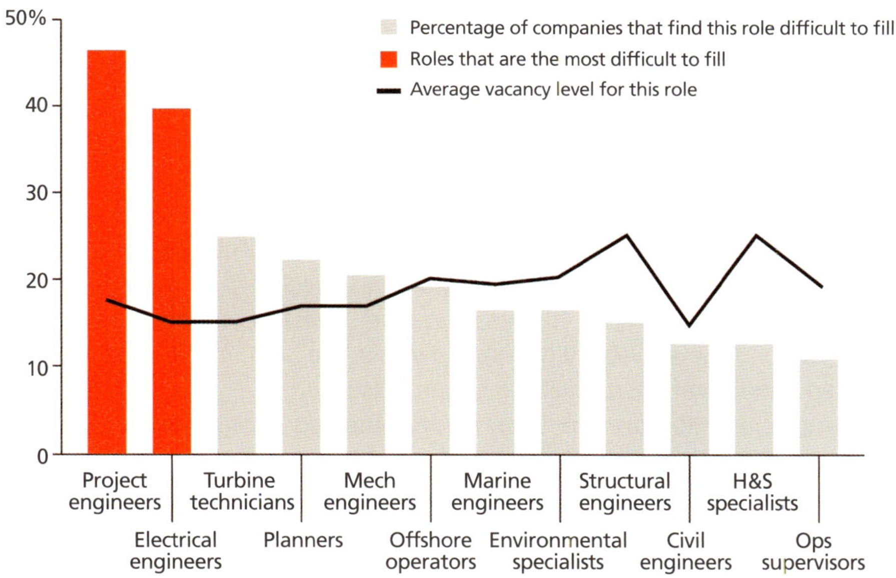

Figure 9.3 UK industry survey of roles particularly difficult to fill
Source: BWEA 2008a

An overview and analysis of key skills important to the wind, wave and tidal energy sector is provided in a report to the BWEA, now Renewables UK (see BWEA, 2008b).

Although some of the issues and barriers highlighted above are likely to be universally observed, the exact extent and importance of these barriers will be extremely country and even regionally specific. For example, countries with existing offshore and gas industries are likely to benefit from the experience and potential transfer of skills between this industry and the offshore renewable energy sector. As another example, countries currently active in shipbuilding activities (South Korea) are extending their offer to the construction of installation vessels for the offshore renewable energy industry, while countries with past or declining shipbuilding industries (the UK and France) are considering the diversification or revival opportunities linked to the development of the offshore renewable sector.

The IEA–RETD REEDUcation project ('Requirements on education and training for a large-scale deployment of renewable energy') will also, once published, provide a further detailed assessment of the barriers and mitigation measures associated with skills.

9.11 Mitigation and removal of skills availability barriers

The first stage towards mitigating the barriers associated with skills shortages requires reviewing the current skills base and the assessment of future requirements at national or regional level. This should be performed through consultation with industry, and based on realistic growth targets for the offshore renewable energy sector and the impact upon local and national job creation based on retention factor.

In a second stage, a strategy should be developed in order to address the skills shortages potentially identified, supported by industry, public and private education providers, and other interested stakeholders. The objectives of any such national skills development programmes are likely to maximize local employment opportunities, increase inward investment opportunities by demonstrating the availability of a skilled workforce, and reduce requirements on international staff recruitment to fill vacancies.

Examples of initiatives which can be implemented in order to remove skills barriers include:

- promotional programmes within the education system – in particular, primary and secondary – to highlight the employment and careers prospects within the offshore renewable energy sector and to encourage uptake in science, engineering and technology subjects in further education or vocational training;
- development of high-quality education and training provision tailored to industry needs;
- promotion of recognizable and trusted qualification standards;

- development of initiatives and financial support packages to assist the retraining and transfer of skills and manpower from other (particularly contracting) sectors of the economy;
- development of a coordinated strategy and action plan between all relevant government agencies, departments or ministries (education, research, business), industry and trade associations, and education providers.

Case study Establishment of National Skills Academies in the UK

National Skills Academies have been initiated by the UK government to address the need for a world-class workforce with better skills, to be employer-led centres of excellence, and to deliver the skills required by each sector of the economy.

Examples of two such academies actively involved in the provision of a skilled workforce for the offshore renewable energy industry include:

- the National Skills Academy for Power (www.power.nsacademy.co.uk); and
- the Energy & Utility Skills (EU Skills) (www.euskills.co.uk).

National Skills Academies are centres of training excellence which have been set up by the UK government in order to support the delivery of skills required in each major sector of the UK economy. Crucially, skills academies are led by employers who work with government and training providers to shape the training and qualifications that will help them to compete in global markets. The National Skills Academy for Power is a recent addition to this network of academies, and is supported by major employers in the UK power sector such as Scottish Power, EDF Energy, Scottish and Southern Energy, E.ON UK, the National Grid, ABB and others.

The Energy & Utility Skills (EU Skills) is the Sector Skills Council (SSC) for the gas, power, waste management and water industries, licensed by the UK government and working under the guidance of the UK Commission for Employment and Skills (UKCES). EU Skills is employer led and its purpose is to ensure that its industries have the skills that they need now and in the future.

9.12 Supply chain and infrastructure barriers

The infrastructure, products and services supply chains supporting the offshore wind, wave and tidal energy industries will need to be vastly developed in order to be able to deliver the ambitious deployment targets set by individual countries (as summarized in Chapter 4).

Table 9.5 summarizes some of the main risks and barriers associated with supply chain and infrastructure, and describes their typical impact.

Supply chain shortages were, for example, one of the contributing factors to the rising capital costs of offshore wind farm projects observed between 2006 and 2010 (see Figure 6.2 in Chapter 6).

Table 9.5 Main supply chain and infrastructure barriers

Barrier	Description and Impact
Product or services supplier shortages	Lack of competition Delays and/or costs increase to project development or cancellation Increased likelihood of poor-quality product or services
Inadequate infrastructure	Delays and/or costs increase to project development or cancellation
Time lags between availability of contractors and project requirements	Delays in project development or construction Contractor loss of confidence, bankruptcy or withdrawal from the market
Macro-economic factors	Global financial crisis, availability of finance Fluctuation in the price of raw materials and commodities (oil, gas, electricity) Exchange rates fluctuation Diversion of resources (finance, staff, vessels, components) to other sectors

The supply chain supporting the offshore wind, wave and tidal energy industries can be mapped according to the following lines:

- product manufacturers and their sub-contractors;
- service providers and their equipment.

Key elements entering the product manufacturer category include:

- technology manufacturers (offshore wind, wave and tidal suppliers and their own supply chains);
- foundation suppliers;
- electrical infrastructure suppliers (offshore substation, transmission cables, onshore substation).

Key elements entering the services provider category include:

- ports;
- onshore and offshore contractors for feasibility, design, construction and operation (including vessels);
- other enablers (including financial institutions, insurance providers, etc.).

Figure 9.4 illustrates an example of a supply chain pyramid applied to wind energy. The key sub-components (blades, generators, etc.) entering into the manufacturing of offshore wind turbines have been presented in section 3.1.3 in Chapter 3. A detailed listing of all major onshore and offshore wind turbine manufacturers and of their suppliers can be found in industry-specialized publications (BTM Consult Aps, 2010).

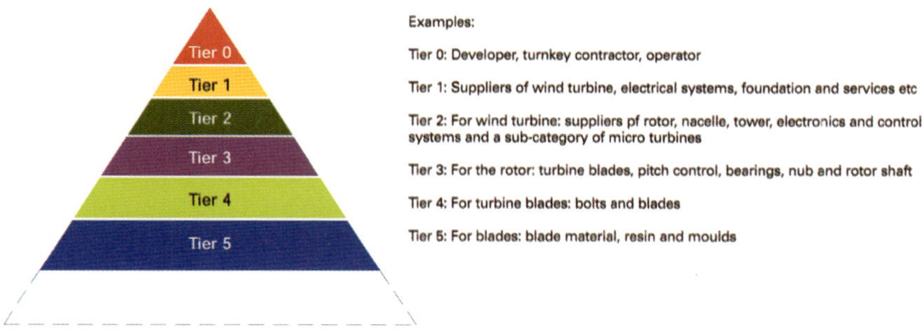

Figure 9.4 Supply chain pyramid: Wind turbine
Source: Renewable Supply Chain Gap Analysis, DTI (2004b)

A good reference describing the whole supply chain of product and services associated with offshore wind farm projects, broken down by stage from early development to operation and maintenance, can be found in a Crown Estate report (The Crown Estate, 2009a).

The requirements associated with grid infrastructure reinforcement have already been covered in section 8.6 of Chapter 8. In the remainder of this section, a brief description of the major supply chain issues associated with each product or service is provided.

9.12.1 Technology manufacturers

There are currently only a very small number of offshore wind manufacturers able to display a proven track record for their technology (see section 3.1.2 in Chapter 3), although this is likely to change in the near future as an increasing number of established existing (onshore) wind turbine manufacturers and dedicated offshore wind technology developers have announced their intention to enter this market and are actively developing their technologies.

Leading offshore wind turbine manufacturers have an established supply chain, and in many cases full control of a large number of their sub-contractors given their vertically integrated structure. In this case, development of their supply chain is mainly associated with increased production capabilities at existing sites and at new sites in response to market demands. Market trends towards increased vertical integration have been observed. A detailed recent description of some observed bottlenecks in the offshore wind turbine component supply chain can be found in a recent report published by The Crown Estate (The Crown Estate, 2009b).

Supply chain engagement is likely to represent a much higher risk factor to small, new offshore technology developers for the following reasons:

- Suppliers may lack confidence in the finance strength or long-term prospects of the technology developer (unless these are, in fact, controlled by much larger companies – for example, offshore wind manufacturer Multibrid is now controlled by Areva and wave technology developer Wavegen is now controlled by Voith Hydro).
- Small initial volumes required are unlikely to be attractive to suppliers unless they consider the strategic advantage associated with an early entry into the sector, or advantage gained by forging early links with the developer (which in a number of cases has led to investment by suppliers into technology developers).
- Small technology developers are likely to lack the resources to manage quality control of their suppliers and sub-contractors.
- Warranty requirements.

9.12.2 Balance of plant

The major supply chain elements included in the balance of plant include electrical substations (offshore and onshore), electrical cables (export transmission cables as well as inter-array cables) and structural support elements (foundations, moorings).

Markets tend to be highly concentrated, which raises the risk associated with shortages. Furthermore, export cables or electrical substation manufacturing, in particular, is highly technical and it will be difficult and require significant investments for new suppliers to enter these markets.

9.12.3 Ports

Offshore wind farm projects require port facilities during construction and later during the operational phase. Some bottlenecks associated with ports are mainly related to:

- the unsuitability of installations in terms of lay-down area, tidal range, capacity, draught, overhead clearance and cranes;
- distance to project development zones;
- distance to turbine and foundation manufacturing production sites.

At a European level, intense competition is now taking place between North Sea- and Channel-facing ports as port owners and authorities have gradually realized the business opportunities created by the expected offshore wind boom. It is, however, likely that only a small number of major port hubs will emerge in the long term.

9.12.4 Vessels

The offshore renewable industry has historically had to rely on installation vessels not specifically designed towards some of its specific requirements. This shortage has been identified for many years; but there was insufficient confidence in the business opportunity provided by offshore wind development for investments in new vessels to be made. The scarcity of suitable vessels, combined with requirements from other industries, has led to full order books for installation vessel owners and operators. A number of offshore wind and tidal projects have experienced delayed implementation due to the unavailability of such vessels at the requested time, or following mismatch between construction schedules and vessel availability.

9.12.5 Contractors and service providers

Experienced offshore construction companies are required to assist project developers and utilities, most of them unfamiliar or lacking internal resources with regard to the construction requirement in the offshore environment. Limitations and issues facing the industry are likely to be linked to skills shortages and to the transfer of knowledge from other sectors (in particular, the oil and gas sector). Traditional offshore contractors have had difficulties in interfacing or managing other contractors and issues related to electrical engineering and grid integration.

The market for operation and maintenance (O&M) services is currently dominated by wind turbine manufacturers as most wind farms are very young and are covered by long-term O&M contracts. A small number of independent O&M service providers are now attempting to enter this market on the back of their onshore wind turbine services track record or offshore (oil and gas) service credentials.

A large number of companies are offering services supporting the feasibility and development stages (surveys, metocean data acquisition, resource assessment, design studies, other studies), with no major particular supply chain issues other than a lack of skilled personnel.

9.12.6 Other enablers

Providers of finance (loan or equity) and insurance are instrumental to the development of the offshore renewable industry sector. To date, only a very small number of financial institutions have been involved in offshore wind project finance deals, and none are likely to consider wave or tidal projects for the foreseeable future. Likewise, only a very small number of insurance firms are currently active in the offshore renewable energy sector.

The critical supporting role of test infrastructures such as EMEC and the Wave Hub in the UK for wave and tidal energy has already been highlighted. A number of new test centres and facilities for offshore wind, wave and tidal technologies (small- or large-scale prototypes and components) are currently under development in many countries. Test facilities, however, require important initial investments and on-going running costs, while facing uncertain revenue streams if these are to be met through fees payable by technology developers. For example, while EMEC's wave and tidal berth sites have been ready for the deployment of prototypes for many years, it is only recently that these have been fully utilized. While currently initiated and mainly financially supported by public funding, a viable long-term economic and financial model for test centres is still to be confirmed.

9.13 Mitigation and removal of supply chain and infrastructure barriers

The main supply chain and infrastructure barriers to offshore renewable energy technologies have been identified as:

- technology manufacturers (for offshore wind: the small number of suppliers) and their supply chain;
- balance of plant;
- ports;
- vessels;
- contractors and service providers;
- others enablers.

The removal of these supply chain and infrastructure barriers will mainly take place through direct investment by the private sector provided that there is sufficient confidence in the long-term market opportunities offered by the offshore renewable energy sector.

The main mitigation measures against the supply chain barriers are clear signals and actions from the governments of countries actively interested in the development of offshore renewable energy projects in their Exclusive Economic Zone (EEZ), and in particular from countries with the largest development potential by capacity. For example, the announcement by the Crown Estate in the UK of the outcome of the Round 3 leasing programme, which corresponds to a potential of 25GW of installed offshore wind capacity, has led to a renewed momentum in the development of the offshore wind supply chain internationally, in general, and in the UK, in particular. Since then, a number of major investments have been announced and inward investments have been secured by the UK from offshore wind turbine manufacturers.

The promotion and development of supply chain capability at a national or regional scale will obviously be of the highest importance for government or local development agencies in order to maximize the economic benefits to be gained from the development of offshore renewable energy projects. These benefits can only materialize if a high retention factor can be secured, particularly for activities and products linked to the construction and installation phase during which the majority of capital expenditures are made.

Economic development agencies and governments have, within the limits imposed by international trade regulations and agreements, the possibility to influence and promote the development of a local supply chain by a number of means. Examples of such initiative include:

- financial support to private companies through grants and investments;
- development and implementation of national infrastructure plans;
- direct stimulation of supply chain by the creation and publication of supply chain directories, marketing documents and other studies, highlighting the market opportunities to existing suppliers and potential entrants.

Case study UK port infrastructure development

Private owners of UK ports have been slow to recognize the opportunities provided by the offshore wind market or undecided in committing the necessary investment. UK and Scottish governments have commissioned a number of report and strategies (e.g. BERR, 2009a; N-RIP, 2010) and engaged with port owners in order to developed a coordinated UK approach to port infrastructure requirements. A specific UK offshore wind ports prospectus (BERR, 2009b) was published in order to promote at national and international level the capabilities and offers provided by UK ports, in a similar way as has been done by German local and national authorities for the offshore wind port and cluster of Bremerhaven.

9.14 Access to capital and financial support mechanism barriers

9.14.1 Historical perspective

Chapter 7 explains the fundamental principles related to offshore renewable energy projects finance.

Wave and tidal energy projects, which are still in the demonstration phase, are not able to raise debt finance from banks. Capital for demonstration projects has tended to come from a combination of balance sheet finance (e.g. from major technology developers) and government grants. Venture capital will also play an increasing role as the technology approaches market. There is a perception that offshore wave and tidal energy lags about ten years behind offshore wind in terms of market readiness and access to finance. This suggests that the first privately financed venture could occur in the second half of the decade.

For early stage offshore wind projects, debt finance has been difficult to access due to the nature of the risk. Instead, balance sheet finance from large corporations or private equity investors seeking to de-risk sites has been applied. A debt to equity ratio of about 70 per cent is required to secure a loan on a more promising project. Subordinate loans to reduce the amount of balance sheet capital are required.

For developed offshore wind projects in the construction phase, combined balance sheet and debt finance have been applied, with support in the form of loan guarantees from export credit agencies and multilateral banks such as the European Investment Bank. Since the first sites were developed, the risk profile of sites has evolved. Levelized cost estimates have increased and yield estimates have decreased, which has broadly raised the risk profile.

At present, from a lender's perspective, the rewards of offshore wind investment generally do not justify the risk. However, lending institutions are pursuing the small number of investment opportunities judged to be viable. The risk of investing in offshore wind development is still perceived to be too high for sources of capital such as private equity, rated bonds, pension funds and bridge loans.

Utility companies continue to be the primary equity investors in offshore wind development.

9.14.2 Risk management and contracting structure

Major infrastructure projects, including those in the power and oil and gas sectors and onshore renewable schemes, are typically implemented using a well-established formula that reduces investor risk and makes projects bankable via a range of investment options, including debt and equity finance.

This formula relies on parties being able to transfer risk down the project line to protect the key parties and limit potential losses. A description of this process is provided in Chapter 6.

In conventional infrastructure projects, lenders and owners manage risk, in part, by taking advantage of a security package which includes a series of contracts, including the engineering procurement and construction (EPC) and operations and maintenance contracts. Under an EPC contract, for example, a single party is responsible for ensuring that design and construction are carried out on time and on budget. If the EPC contractor fails to deliver the project to budget, the EPC contractor must absorb the loss, and if the project is not delivered on time, the lender is protected, at least up to a given point through the applicability of 'liquidated damages' clauses that require the EPC contractor to pay for the resulting loss in revenue. Provided that the EPC contractor has a strong balance sheet or adequately manages its own risks, this mechanism protects the lender from the financial impacts of project development.

An EPC contract also removes one of the most significant risks of large-scale project development, which is the risk of multiple interfaces between different contractors and suppliers. The responsibility and risk of managing these interfaces is transferred to the EPC contractor.

The EPC contractor must, in turn, manage its risks through effective project management and contractual mechanisms that pass down liquidated damage clauses to contractors and require product guarantees from suppliers.

One could say that from a project risk standpoint, offshore wind is roughly a decade behind onshore renewables and, at present, few entities are willing to act as an EPC contractor at a reasonable price due to the risk premium. Liquidated damage clauses are also not an option in the present risk environment. The effect of this is that the lender and owner are left holding a much higher percentage of the risk than more established projects in the thermal power, onshore renewables, and oil and gas sectors. However, a multi-contract approach is also often used onshore, offshore and in other technologies. Interface risks can be managed.

9.15 Mitigation and removal of financial barriers

Financial barriers can be encountered at each phase of technology commercialization, from R&D to market entry and commercialization.

Government-backed fiscal mechanisms for overcoming barriers to finance in developing and deploying marine renewables can be categorized as:

- expenditure support measures (capital grants, soft loans, tax allowances, accelerated depreciation);
- income support measures (feed-in tariffs, renewable quotas, green premiums, tradable green certificates).

A summary of the application of these fiscal and other support mechanisms is outlined in Table 9.6.

Governments can also back a range of measures to reduce the risk of technology deployment and attract investment, which is covered in other chapters of this book.

Typically, expenditure support measures (market 'push') are more prevalent for technology development and early stage deployment, while income support measures become the main support mechanism as technologies move into larger-scale 'commercial' deployment (market 'pull').

Table 9.6 Summary of potential mitigation measures barriers to finance

Barriers	Potential Mitigation Measures	Examples
Early stage technology research and development with no near-term commercial prospects	Government grants, tax incentives for R&D	See section 8.6.1
Lack of business planning skills	Capacity-building, incubator programmes	
Late stager development: unproven technology	Government grants for demonstration Funding for test centres and validation, incentives for government and private-sector collaboration	See section 8.6.3
Inability to gain finance for scaling up due to insufficient cash flow	Capital grants, public procurement mechanisms, incentives for the adoption of innovative technology (tax incentives), soft loans, loan guarantees	
Market failure due to high cost of new technology	Income support measures, regulatory drivers to adopt new technology, strategies to develop economy of scale	
Project development risk	Loan guarantees, streamlined permitting, site pre-assessment (SEAs)	
Project interface risk	Tendering arrangements that reduce the number of contracts, especially at key interfaces	Allowing generation developers to develop offshore transmission networks (UK)

A summary of some expenditure support mechanisms is provided in Table 9.7 (IEA, 2010).

Table 9.7 Expenditure support mechanisms

Country	Tax Incentives	Grants	Loans/Loan Guarantee
Belgium*	Tax deduction for investments in renewable energy by enterprises (1992): 13.5% tax deductions for renewable investments		
Canada	Accelerated capital cost allowance (2007): allows accelerated write-off of renewable energy assets Canadian renewable conservation expenses (1996): allows corporations to class expenditures associated with start-up of renewable projects as fully tax deductible		
China	Preferential tax policies for renewable energy (2003): income tax on foreign investment into wind projects reduced from 33% to 15% Reduced VAT for renewable energy (2001): VAT for wind power cut from 17% to 8.5%		
Denmark*		Subsidies for wind turbines and other renewable energy generating plants (IEA, 2009a)	

Table 9.7 continued

Country	Tax Incentives	Grants	Loans/Loan Guarantee
Finland*	Tax refund of €6.9/MWh for wind (IEA, 2009a)	Investment subsidy up to 40% depending upon novelty of system for wind (IEA, 2009a)	
France*	Flexible depreciation (2003): allows accelerated write-off of renewable energy assets		Government crediting and loan guarantee for renewable energy investment (2001): covers up to 70% of loans from banks to SMEs in renewable activities
Germany*		Integrated Climate Change and Energy Programme (2008): contains subsidies for offshore wind farm development	KfW Renewable Energies Programme (2009): offers preferential loans and grants for renewable projects
Ireland*	Tax relief on corporate investment (Dalton et al, 2009)		

Country	Tax Incentives	Grants	Loans/Loan Guarantee
Italy*		€14 million funding for various projects including renewables (2010)	
Japan			
The Netherlands*	Energy investment deduction (1997): allows up to 44% of renewable energy investment to be deducted from taxable profit	SDE production subsidy paid as a tariff on the basis of average production cost	
Norway	CO_2 Tax (1991): tax on combustibles and emissions from the petroleum industry	Investment subsidy for demonstration projects	
Portugal*		35% subsidy on investment up to €2.5 million and SIEST programme of 25% subsidy on technologies within Azores Region	100% reduction on interest of loans up to €750,000
Spain*			
Sweden*	Energy, carbon and SO_2 taxation (1991): complex taxation of CO_2, NO_x and sulphur emissions		
Taiwan			

Table 9.7 continued

Country	Tax Incentives	Grants	Loans/Loan Guarantee
UK*		UK£1 billion upgrade in grid network (E&Y, 2010) Direct support in manufacturing facilities (E&Y, 2010)	Plans to launch UK£2 billion 'green investment bank' fund
US	American recovery and reinvestment act: tax-based provisions (2009): allows a 50% write-down on renewable projects; contains various tax credits for renewable energy technologies Modified accelerated cost-recovery system (2008): allows recovery of investment into wind technology to be depreciated over accelerated timescales Production tax credit (PTC) and investment tax credit (ITC) (E&Y, 2010)	Option to convert PTCs and ITCs into treasury grants (E&Y, 2010)	American recovery and reinvestment act: tax-based provisions (2009) – offers clean renewable energy bonds DOE Loan Guarantee Programme (2007): guarantees loans for renewable projects

Note: * Subject to EU 20-20-20 climate goals by 2020: 20% cut in 1990 emissions, 20% increase in share of renewables in the energy mix, 20% cut in energy consumption (European Union Climate Action, 2010)

The European Commission also provides grants, such as the €110 million for an electricity interconnector between Ireland and Wales to facilitate renewable energy deployment (E&Y, 2010). A list of existing and planned European test facilities can be found in Dalton et al (2009). A summary of income support measures is shown in Table 9.8.

Table 9.8 Income support measures

Instrument	Pros	Cons
Feed-in tariff	Provides price certainty over long period Can be differentiated by technology/location Administration costs are low Easier to adjust to address changes in cost base than market-based systems	If provided for limited time or prices subject to review, can create 'boom and bust' development scenarios
Quota obligation	Provides volume certainty Incentives can be coupled with compliance costs or penalties Tradable green certificates can provide additional income support	With financial compensation, all obligated parties have to count on recovering compliance costs in wholesale market Provides no price certainty for investors
Certificate systems	If price is floating, then can be economically efficient in delivering volume for least cost	Provides no price certainty for investors Can require complex trading systems and mechanism to protect against creditor default Tends to be top-up to grey tariff so that investors are still exposed to price risk
Fiscal incentive (green tax exemptions)	Easy to administer	Tends to provide limited financial support (i.e. just top-up) so that main benefit rests on grey tariff
Tendering arrangement	In theory should provide least cost solution to achieve given volume Can be banded by technology Provides price certainty to investor	Requires insightful screening process (hence high admin load) in order to sift out unrealistic offers Implies picking winners

Suppliers of debt finance, such as banks, tend to favour feed-in tariffs due to their long-term certainty. Balance sheet investors, such as utility companies, tend to favour quota systems such as renewable obligation certificates because additional value can potentially be achieved.

The underpinning stability and sustainability, including allocation of costs for financial instruments, also play an important role. In a number of countries, feed-in-tariff costs are passed on to the government, even if they are initially paid by power suppliers. This is the situation in Spain, which recently reduced feed-in-tariff pay-outs for some technologies in response to economic conditions. In the case of quota obligations, the ability of a utility to pass on its compliance costs to consumers may or may not be restricted by regulators.

In general, the level of support for marine renewables tends to be in line with wind and bioenergy support and less than that for PV. In fact, few jurisdictions make any special allowance for the higher costs of these immature technologies, though differentiation is increasingly being applied to offshore wind as opposed to onshore wind.

9.15.1 Comparison of support value with levelized costs

We have compared the value of support with the levelized costs in different countries in order to assess the current effectiveness of the market pull that these measures are trying to create. Assuming that projects start today, we have estimated the value of revenues, including income support measures, over the life of the project versus the projected levelized cost to provide an indication of underlying economic attractiveness expressed as a net present value (€/MWh over life of project). Where feed-in tariffs have a shorter duration than the project life, we have assumed that project revenues will fall into line with projected market prices. The same projected market prices also provide the base income to which green certificates/green premiums are added.

For wholesale prices we assumed that wholesale market prices are 90 per cent of large industrial power prices as reported by the EU energy portal and then escalated at the real growth rates for average US wholesale electricity prices as projected by the US Energy Information Administration in its 2010 *Long Term Energy Outlook*. This shows a 13.4 per cent increase over 30 years. As a sensitivity case, we have also tested the results versus the levelized cost of a new gas-fired combined cycle gas turbine (CCGT) operating at a 70 per cent annual capacity factor, which could be reasonably assumed to approximate long-run baseload energy prices in most jurisdictions. This cost of CCGT plant works out at about €75/MWh on the basis of a US$7.5/mmBTU gas price and €20/tonne carbon price. The other assumptions are shown in Table 9.9. Gas-fired CCGT is not the appropriate benchmark for China; however, our estimated coal generation levelized cost for China would be close to this level, assuming Chinese EPC costs.

Table 9.9 Combined cycle gas turbine (CCGT) cost build-up

Items	Value
Equipment (cost: US$/kW)	900
Logistics/delivery add-on	25%
Annualized investment @7.5% discount, 25 years: US$/kW/year	80.7
Annual capacity factor	70%
Capital cost: US$/MWh	13.2
Annual OPEX (ex-fuel): US$/kW/year	27.0
Annual OPEX (ex-fuel): US$/MWh	4.4
Variable non-fuel cost: US$/MWh	1.5
US$ million BTU gross	7.5
Efficiency %: HHV	52%
Fuel generation cost: US$/MWh	68.1
Carbon price: US$/tonnes CO_2	26.0
Carbon cost: US$/MWh	10.1
All-in cost: US$/MWh	97.3
All-in cost: €/MWh (@1.3)	74.9

Levelized cost estimates for marine renewables have been taken from the analysis undertaken in Chapter 5 of this book. For simplicity, we have assumed the same range for all countries: €120 to €250/MWh. The low end of the range corresponds to shallow water, near-shore and high-energy yield sites, while the high end reflects lower yields and tougher operating conditions.

The results of the net value of income less levelized costs are reported in net present values (NPVs), which take account of the changing support arrangement over time. Positive values indicate that the financial support framework provides a level of support that would allow recovery of the project's levelized costs. Negative values indicate a shortfall. These figures are on a pre-tax basis. Japan is not included as no financial support mechanism is currently in place for offshore technologies. Results are reported in Figures 9.5, 9.6 and 9.7, respectively, for offshore wind, wave and tidal energy projects.

The results clearly indicate that in most national jurisdictions, the level of support is inadequate to cover levelized costs, even at the low end of cost projections, assuming differentiated wholesale prices. Only in four countries

(Italy, France, Belgium and the UK) are the support levels sufficient to cover the low-end levelized cost projections across the three technologies. Denmark and the Netherlands have positive returns for offshore wind. This result holds at both a 3 and a 10 per cent real discount rate.

However, in no cases is national support sufficient to cover the higher-end levelized cost cases. Figures 9.5, 9.6 and 9.7 show the NPV values of revenue versus levelized costs based on existing income support measures and the differentiated wholesale prices. The countries with full feed-in tariffs see no difference; however, for those with income linked to market prices, there are small differences. On balance, the returns are slightly reduced versus the case of differentiated wholesale prices.

It should be noted that in the real world, investors will tend to discount very strongly incomes which are uncertain, such as market-based support instruments, as opposed to fixed feed-in tariffs. The returns are shown to be highest in the UK. However, even here investors complain that such projects are not viable given the risk discounts applied to the value of market-based support instruments.

The analysis also assumes no tax liabilities. In practice, corporate taxes would reduce the post-tax returns. Clearly, favourable tax treatment of capital expenditure (CAPEX) spend can reduce the tax liabilities (and increase project internal rate of returns (IRRs) versus baseline taxes); however, post-tax returns will always be less than pre-tax returns.

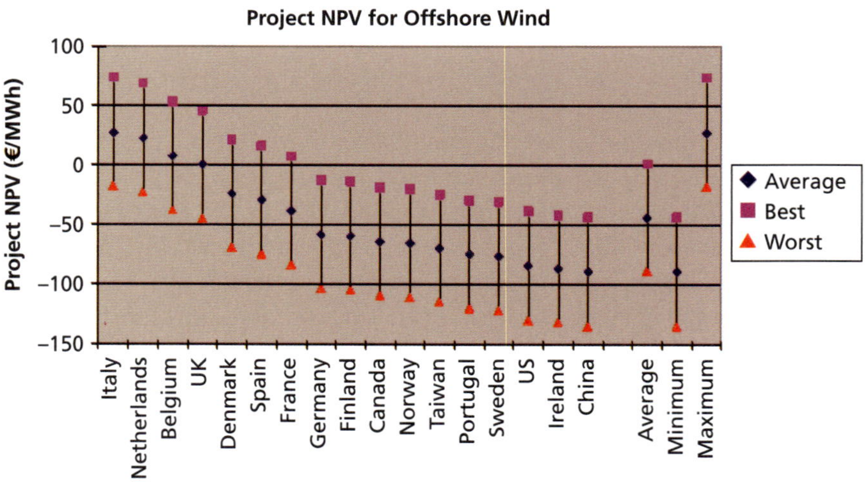

Figure 9.5 Offshore wind project economic NPV in €/MWh based on existing income support schemes under a high and low levelized cost projection: Differentiated wholesale prices

Source: Mott MacDonald

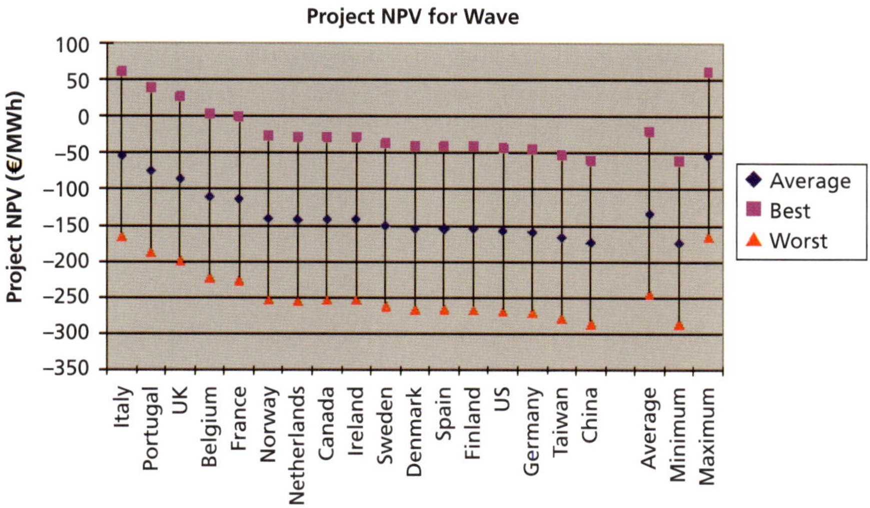

Figure 9.6 Wave energy project economic NPV in €/MWh based on existing income support schemes under a high and low levelized cost projection: Differentiated wholesale prices

Source: Mott MacDonald

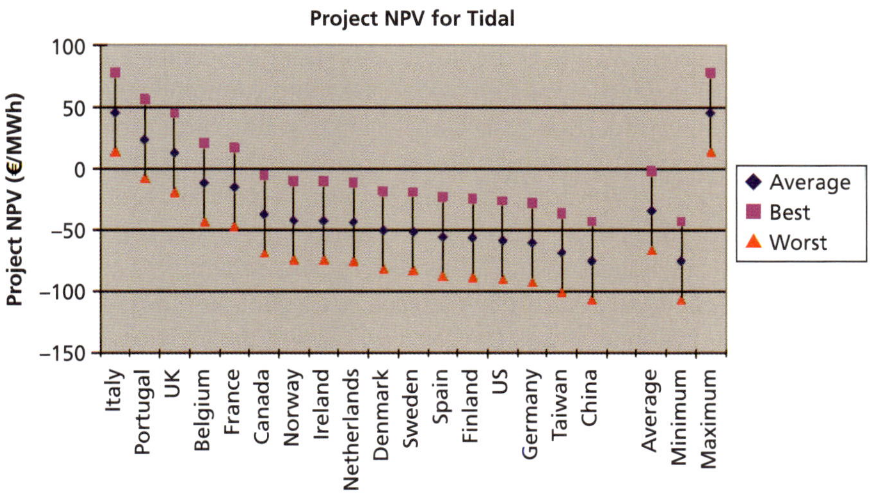

Figure 9.7 Tidal energy project economic NPV in €/MWh based on existing income support schemes under a high and low levelized cost projection: Differentiated wholesale prices

Source: Mott MacDonald

These calculated returns do not include any additional contribution from sale of carbon offsets. In principle, renewable generators would be eligible for carbon offset income, either through the Joint Implementation, Clean Development Mechanism (CDM) in Non-Annex I countries, emission reduction units under the

European Union Emissions Trading System (EU ETS) and voluntary or regional trading schemes, in addition to their national income support measures. However, there are few cases of this being applied, primarily for two reasons. First, the value of the carbon reductions generated is small in comparison to the financial needs of the project. Second, the presence of higher-value financial instruments, such as feed-in-tariffs and other government drivers, makes it more difficult to demonstrate additionality, as required by regimes such as the CDM.

However, the long-term uncertainty and risk associated with carbon prices, emission allowances and compliance with future climate change regulations does have a bearing on investment in offshore wind, especially by utility companies. These and other less tangible financial factors are discussed in the following section.

9.15.2 Motives for offshore renewables investment

There are additional motives for investing than pure economic considerations based on project financial returns on a traditional discounted cash flow analysis. These tend to be categorized as strategic motives.

Large portfolio generators may take the view that there is a high degree of uncertainty regarding future fuel and carbon prices, future emission standards and plant life extensions, and the ability of investors to bring on new nuclear plant, and that having a certain amount of centralized renewable generation capacity available provides a valuable hedge. Offshore wind energy is seen as offering new prospects for large-scale deployment of wind energy projects, as in some European regions opportunities for further onshore deployment are becoming limited. Such investors often take the view that the first projects need not achieve hurdle rates if the project allows significant learning and therefore capitalize large cost reductions and better IRRs for subsequent projects. Where large portfolio generators have downstream retail arms which give them a significant market position, they could anyway recover the additional costs of new renewable projects from their retail customer base by keeping prices higher than otherwise.

Major utilities may also take the view that they will have a reasonable chance of persuading governments/regulators to improve or extend income support measures, or take other measures to reduce development. These companies have considerable bargaining power (the power to decelerate/accelerate investment) in negotiations with governments keen to meet their renewables targets.

A number of utilities have also taken a wider strategic position by actually investing in the supply chain associated with offshore renewable energy technologies. For example, Scottish & Southern Energy have invested in BiFab (an offshore wind turbine foundation manufacturer) and Aquamarine Power (a wave technology developer), while SP/Iberdrola have invested in Gamesa, a wind turbine manufacturer which has recently announced its intention to enter the offshore wind market. Some of the rationale behind these investments is to remove some supply chain risk and have better control over costs. It may also provide an opportunity to later sell their interest.

Strategic motives are more applicable to large utilities which can finance projects on balance sheet rather than independent developers who would rely on project finance and for which commercial rates of return will be required.

9.16 Conclusions

Offshore renewable energy technologies and projects face a large number of non-technical barriers and challenges to their development and deployment. The main general barriers and challenges are linked to environmental, safety, regulatory and licensing, competing uses, skills, supply chain and infrastructure, and financial issues. While all of these issues are important and have to be dealt with, mitigation measures can reduce their impact.

The main mitigation measures in order to ensure the minimization of environmental barriers and acceptance from other sea users are early engagement with all stakeholders, appropriate marine spatial planning and adoption of the recommendations from the environmental impact assessment (EIA).

The reduction or mitigation of health-and-safety barriers can be achieved by a strong industry culture, supported by staff training, compliance with legislative requirements, best practices and standards, as well as through technical innovations.

Complex permitting processes are another major barrier to offshore renewable energy project development in most countries. Prescriptive planning conditions or requirements limit projects and technology design options and can significantly increase timescale and development costs. A number of regulatory barriers are also delaying or preventing the changes required in the onshore and offshore grid infrastructure in order to accommodate offshore (and onshore) renewable expansion plans. Deployment timescales can be greatly increased as a result of these barriers.

While the permitting processes are diverse and country specific, lessons can be learned from the countries that have had more success with offshore wind. Streamlined application procedures, one-stop shops and pre-permitted areas are some of the potential mitigation measures to planning and permitting barriers. The allocation of seabed rights to competent and construction-focused developers is also important in order to prevent sites from being leased to developers more interested in speculative applications or without the necessary resources to progress the development of projects.

A large number of anthropogenic constraints are faced by project developers as a result of current or future use of open water spaces for commercial, recreational or military activities. The most appropriate mitigation measure against barriers associated with competing use is the development of marine spatial plans involving, and in consultation with, all relevant stakeholders at local, regional, national and international levels.

The removal of skills barriers requires the active promotion of the various employment and career opportunities provided by the offshore renewable energy industry, as well as the development of training courses and programmes tailored to the needs of the industry.

The infrastructure, products and services supply chains need to be vastly developed in order to increase competition and avoid shortages. This can be delivered by private-sector investment, but only if governments establish sufficient confidence in the long-term market opportunities.

Access to capital and financial barriers can be encountered at each phase of technology commercialization, from R&D to market entry and commercialization. These are particularly important for wave and tidal technologies, which for the most advanced technologies are typically at a pre-commercial/prototype stage, and consequently are seen by financiers and investors as containing large amounts of technology and performance risks.

The lack of long-term or stable policy commitments from governments is another significant barrier as it affects developer and market confidence. Furthermore, in many countries, the level of financial support provided (feed-in tariff or tradable certificates) often appears either insufficient or, at best, marginal in order to provide sufficiently attractive returns to investors compared to lower risk investment options in other sectors. Government-backed fiscal mechanisms must be offered for overcoming barriers to finance in developing and deploying marine renewable technologies, such as expenditure support measures (capital grants, soft loans, tax allowances, accelerated depreciation) and income support measures (feed-in tariffs, renewable quotas, green premiums or tradable green certificates).

Notes

1 See, for example, SMRU Limited (2010) for best practices on marine mammal monitoring.
2 Note that a new law has been passed by the Norwegian Parliament in March 2010, providing the legal framework for resource allocation for new energy production in Norwegian waters.
3 NRG Bluewater Wind (2010).
4 See http://www.dnrec.delaware.gov/News/Pages/Delaware_becomes_first_state_to_receive_delegation_from_EPA_for_offshore_wind_permitting.aspx.

10

Guidelines for Project Development

This chapter provides a set of guidelines for the development of offshore renewable energy projects. Figure 10.1 illustrates the main stages during the lifecycle of an offshore renewable energy project.

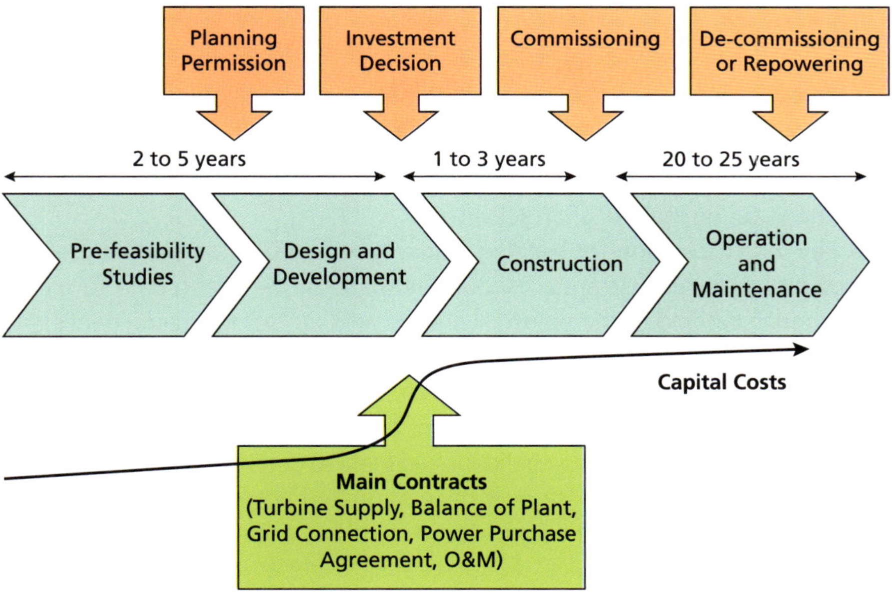

Figure 10.1 Typical offshore renewable energy (offshore wind) project lifecycle

10.1 Stage A: Opportunity analysis

10.1.1 Project development strategy

Project objectives should be defined before any other work commences to reflect the developer's aspirations. Project objectives will be referred to during subsequent stages of the project development in order to decide whether the objectives are being met and whether to continue with the project.

An outline project development strategy should be developed to discuss how the identified objectives will be met. A market analysis should be undertaken to identify competitors and clients, as well as to determine financial, environmental and legislative constraints. The market analysis should highlight the key opportunities and risks associated with developing the project further.

It is at this early stage that key project parameters should be outlined and a project risk register should be created. The risk register should be kept as a live document throughout the project life as it will act as a useful tool to record and then address the key project risks. The risk register will also help to drive the project forward and will allow the development to be continually assessed in terms of technical and commercial risk.

10.1.2 Site screening (pre-feasibility)

A desktop screening exercise should be performed, based on available data, and will identify one or more suitable locations for the project within the area being considered. The site screening exercise will focus on criteria such as port access, proximity of sensitive receptors, energy resource availability and grid connection, as detailed below. The screening exercise will result in a shortlist of candidate sites.

A nearby port with the appropriate facilities should be identified to confirm that there are sufficient loading and storage facilities for the purposes of laydown of equipment during construction and from which to launch any equipment during maintenance activities.

A key search criterion for suitable marine sites will be an assessment of the energy resource. An assessment should be made of metocean data to gain a better understanding of the potential energy yield, access for maintenance and general conditions that any sea-borne device will be expected to endure. A review of seabed morphology and the water depth should be undertaken to ensure that they coincide with the needs of the technology.

Grid connection points must be identified early to determine whether there is adequate capacity available for new generation and how much upgrade work may be necessary. Distance to grid connection has a major impact upon the capital cost of offshore devices as the cost of cable and laying operations is significant.

The project development process should identify whether a strategic environmental assessment (SEA) has been developed for marine power projects in the area that could be used to focus the size of the search area. If not already covered within a SEA, regard must be given to the relevant planning and legislation requirements, along with jurisdictional limits and restrictions affecting the site; these may or may not feature as risks in the risk register. A search for designated protected areas should be undertaken to account for both local and international requirements. A study into local protected species and their level of sensitivity should be undertaken in order to evaluate the potential ecological impact at the nominated site. In addition, a search for archaeological and historical sites should be conducted.

Other sea users and infrastructure should be identified, along with the key statutory consultees and the local coastal planning authorities. Stakeholder analysis will enable the project development team to initiate early (informal) discussions with other stakeholders in order to create an open and smoother development process that encourages knowledge sharing from local people.

Any health-and-safety hazards that are identified at this point should be recorded on the risk register for future consideration.

Preliminary financial evaluation should take place.

10.1.3 Project feasibility analysis

Project feasibility should commence once one or more sites have been short-listed. In this phase, more detailed investigations should be undertaken in order to prepare a conceptual design and technical specification for the project. Key issues to be considered are discussed below.

Regulatory environment

A comprehensive list of consultees and stakeholders should be made, which will lead to formal consultation and discussions with those identified. The local planning authority should also be consulted to assist in identifying potential constraints. Preliminary discussions should be initiated with the owners of onshore and offshore sites likely to be affected by a development, and contact should be made to inform the local community of the project and its impacts.

Resource and site conditions

Marine resource, onshore ground conditions, energy resource availability, bathymetry and seabed geomorphology all need to be investigated in further detail in order to gain a thorough appreciation of the technical and physical issues.

Technical solution

A review of different marine energy converters should be performed based upon the site-specific constraints and the operational history of the devices. A shortlist should be created and the final selection should be made following a high-level financial modelling exercise.

Power export from site

Technical options for grid connection should be assessed and discussed with the operator of the grid network in order to identify the costs of any future energy distribution.

Environmental, social, health and safety

An environmental scoping study should be prepared and submitted to the consenting authorities on the environmental issues highlighted during the site screening. The relevant leasing organizations should also be identified and contact should be initiated.

An appropriate Health and Safety Management Plan (HSMP) should be drafted to identify hazards and to clearly assign responsibility for managing the risks. The HSMP should be continued as a live document and reviewed and updated throughout the project and will allow a review of risk management performance. Special attention should be paid to the risk of severe weather conditions experienced during offshore operations. According to legal requirements, a competent designer should be appointed at this stage.

Constructability

Construction methodologies, logistics, vessels and ports should be assessed.

Schedule

Realistic timescales, including provision for weather downtime, should be estimated.

Costs

Estimation of capital, operational costs and decommissioning costs should be performed in order to identify the key programme constraints. The required insurance and taxes for each stage of the project should be considered and included in both the project life costing estimates and the cash-flow model.

Revenues

The developer should begin negotiations with electricity suppliers and address plans for power purchase. The power purchase agreement (PPA) can be arranged based on the appetite that the developer has for risk associated with power price, and different forms of PPA include fixed-price contracts and floor price contracts. The main differences between the two are that the floor price contract offers a possibility of higher return rates through higher price of electricity, but the floor price may end up being lower than the fixed-price option. Similar negotiations should be held regarding the sale of renewable obligation certificates (ROCs), which can be sold for an upfront payment or payment delayed until the end of the year when the ROC recycling price may offer a higher reward.

Financing model

A financial assessment should be performed to facilitate the creation of a simple financial model showing an initial appreciation of the project economics. The financial modelling will be based upon initial energy yield predictions, predicted revenue streams, availability of external funding, and indicative costs (installation and operation and maintenance).

Business model

Commercial aspects of the developer's business should be addressed, such as production of a business model that is specific to the project objectives. A business cash-flow model should be derived that will present the return on investment at all stages. Plans regarding handover and operation should be developed and a comparison should be made of different development routes, ownership routes and sources of finance. Long-term economic support should be considered to determine the bankability of the business.

Risks and risk management

A risk matrix should be developed and mitigation measures assessed.

10.2 Stage B: Project materialization

10.2.1 Design

If the feasibility assessment meets the project objectives and no prohibitive environmental issues have been identified, the project developer may progress to design.

Consent should be gained from consenting authorities at this stage. The studies undertaken during concept feasibility should be used to refine the project design and, similarly, the technical information from the design should be fed back into the consent application. A procurement strategy that best fits project objectives should be defined in parallel with the negotiations for a land lease agreement.

An environmental impact assessment (EIA), the content of which should have been defined in the environmental scoping study, should be undertaken as a systematic and stepped process consisting of the following three stages:

- surveys and specialist investigations;
- modelling and specialist studies to the standards specified in the scoping opinion; and
- inputs from consultees from continued dialogue on the scope of surveys and studies, likely impacts and mitigation measures.

The environmental study should be made available to the general public for the purpose of post-consent application, which is very important since the consenting authorities will be asking for the opinions of people who may be affected by the project. A public exhibition and presentation to the stakeholders should be organized to inform them of the results of the EIA.

Other environmental parameters should be assessed in the form of an environmental statement, which should demonstrate to the consenting authorities that all reasonable measures have been taken to minimize the environmental impact of the project during its lifetime. The environmental statement should include as a minimum:

- project description;
- EIA process and methodology;
- planning and policy framework;
- summary of environmental assessment;
- monitoring requirements.

A non-technical summary of the environmental statement should also be prepared for wider public distribution. The consenting authorities will then make a decision on granting consent; the length of this process will vary between projects and between countries.

In parallel to the EIA, a project design should be developed that will set the basis for suitable procurement and contract strategies. The procurement strategy should meet the project objectives and risks. The key factors that need to be considered when designing a procurement strategy and a contract form are:

- elements of the project to be procured;
- current market status and projected market trends;
- rules and procedures for procurement applicable to the project development;
- contract management experience of the project developer;
- pricing strategy;
- risk allocation and management (special attention should be paid here); and
- procurement process timescales and integration with the overall project programme.

Health-and-safety issues should be recorded at all times.

At the conclusion of the design stage, and provided that consent and investment decision has been secured, the project can progress to the procurement and construction phase.

10.2.2 Fabrication and installation

If the technical and commercial performance of the project remains in line with the project objectives, the stage of detailed design, project fabrication and installation should be undertaken upon receipt of the consent.

Various technical studies must be conducted to refine the design, which should be developed in accordance with offshore standards. A quality plan should be prepared to specify the required standards.

Tender documents should be prepared that clearly define the scope of work, conditions of contract, terms of payment and risk allocation. As a minimum, tender documents should contain a project description, technical specifications and key conditions of the contract. An assessment of the suitability of the tenders should then be undertaken and procurement contracts signed provided that the conditions meet the previously documented procurement requirements.

Once the contract(s) have been awarded, project fabrication can start. The quality plan and the health-and-safety risk assessment should be used to monitor manufacturing activities according to the standards, timescales and costs agreed. Factory acceptance testing should also be carried out to confirm that the products comply with the project specifications.

A technical representative should be appointed to supervise the contractor activities during the project installation. The scope of works of the installation contractor should include the implementation of the environmental provisions, and environmental monitoring should be undertaken during the installation. Where there is more than one contractor, a principal contractor should be appointed to manage the site. The principal contractor will have prepared a detailed HSMP during the tendering process that will be used during the construction phase. Testing must be undertaken at the end of the commissioning, before ownership is taken by the operating organization.

The licence conditions should include reference to hazard identification and mitigation. The extent of insurance during this phase should be clarified and the installation methodologies should be prepared prior to any activity.

During the construction work, consultation and liaison with other stakeholders should be maintained in order to minimize disruption to local residents and maritime users. Communication protocols should be in place to inform stakeholders of the different installation tasks and likely impacts. A liaison officer should be appointed to ensure communication between the various affected parties and to maintain contact with local planning authorities. The liaison officer will provide support in case of an incident.

Environmental monitoring and management should be performed according to the conditions given in the permit and in order to comply with legislative requirements.

This phase concludes with the commissioning of the offshore renewable energy devices (offshore wind turbines/wave or tidal converters, and associated electrical infrastructure).

10.3 Stage C: Reliability and sustainability

10.3.1 Operation and maintenance

Safe and cost-effective operation should be ensured throughout the lifetime of the project by the preparation of a thorough operation and maintenance (O&M) strategy. Suitable maintenance should be scheduled and health-and-safety measures should also be planned and implemented. Procedures for operating and maintaining any plant throughout the life of the project should be prepared in advance of commissioning and the installation contractors should give technical support at the early stages of operation (this should be included as a condition in the procurement contract).

An O&M plan should be prepared and reviewed over the life of the project that will set performance parameters against which the operation will be monitored. A HSMP should also be prepared and implemented to ensure that hazards are identified and risks mitigated effectively throughout the life of the project. Special attention should be given to emergency procedures during operation and maintenance of the site.

Consultation and liaison must be maintained to ensure minimum disruption to the local environment and community. Liaison is also important in order to avoid disruption to project activities by other activities undertaken in the area. Adequate communication protocols should be put in place, along with effective emergency response procedures.

10.3.2 Decommissioning

Decommissioning plans should be established to facilitate the safe, cost-effective and environmentally sustainable decommissioning of project infrastructure at the end of the project's life. A plan for decommissioning should be prepared as part of the consent conditions and revised over the life of the project. A fund to cover the cost of decommissioning should be set aside during the project life to ensure that adequate provision is made, even in the case of insolvency. A procurement strategy should be designed for the outsourcing of the decommissioning work.

Before the work commences, a safe means of recovery, temporary storage and disposal of equipment must be set up and the procedures for the decommissioning work will be laid out in detail. Approval must be obtained before the start of the work.

The HSMP will require adaptation to cover the decommissioning period and special attention should be paid to severe offshore working conditions.

Environmental impacts of decommissioning must also be identified and mitigated and the same consultation process adhered to as was implemented at the construction stage.

10.4 Conclusions

Developers are recommended to follow best practice at all stages of a project's lifecycle, from pre-feasibility, development, design, construction, operation and decommissioning. Examples of such best practices include definition of clear project objectives and development strategy, early consultation with other stakeholders, and creation of a risk register to be maintained as a live document throughout the project life as a tool to record and then address the key project risks.

11 Findings, Recommendations and Model Policy Framework

11.1 Findings

11.1.1 Industry context

The world theoretical resource from offshore renewables (wind, wave and tidal) is estimated to be between 260,000 and 330,000TWh/year, illustrating the potential significance of the available resource. The opportunity to harvest this vast resource has been identified by governments and academia, together with commercial project and technology developers, who aim to capitalize in a rapidly expanding market.

Of all the marine technologies, offshore wind is the frontrunner, with commercial projects operating since the early 1990s. Even though the offshore wind sector is experiencing high growth, the industry is far from mature and big challenges lie ahead, with projects being planned for deployment further offshore, in deeper locations, with larger machines and technological advancements.

For the wave and tidal sector, a large number of devices are under development, with no particular design having yet emerged as clear frontrunner. The various technologies are at different stages of development, with some prototypes currently being tested at full scale and commercial projects expected in the near future.

Due to the harsh and difficult-to-access environment in which these devices have to be installed and operate, the associated risks (technical and non-technical) are higher than for onshore technologies. Complete removal of such risks is not feasible; however, mitigation measures can reduce these risks to an acceptable level to facilitate project development.

11.1.2 Costs

Regardless, however, of the mitigation measures, the costs of marine projects remain high and uncertain, resulting in financing of a marine project being the biggest barrier for their deployment.

Offshore wind is currently the cheapest of the marine technologies in terms of cost of energy for an installed project (including transmission connections to the shore) with a range of €120 to €250/MWh. Initial estimates of cost of energy for wave projects are in the region of €140 to €530/MWh, whereas for tidal projects these costs are in the region of €110 to €220/MWh installed. Cost of energy for commercial wave and tidal devices are, however, initial estimates and should be treated carefully as the uncertainty inherent in these projections is very high.

11.1.3 Financing

There are a number of financing options available for projects developed by the private sector which are primarily balance sheet and project finance. Each option has its benefits and drawbacks. The characteristics of individual projects and their sponsoring organizations typically dictate which one of these financing options is best suited for a particular project. Balance sheet finance using debt raised corporately is cheaper, involves fewer parties, and control of the project remains firmly with the owner; it is, however, capital intensive and the risk of failure lies entirely with the owner. On the other hand, project finance allows greater leverage from the available funds for sponsors' equity investment; however, it is typically more expensive and complex and an element of control over the project is afforded to the lenders. One solution is to finance construction projects on balance sheet and move to project finance on completion, recycling development capital into new projects.

To date, wave and tidal stream projects have not been project financed. With the most advanced technologies typically at a pre-commercial/prototype stage, they are seen as containing large amounts of technology and performance risks. Funding for technology deployment, to date, has tended to tap venture capital or public-sector development support sources. Project developments are mainly pursued by utilities. A project finance model may emerge in the future once the technologies have been de-risked.

11.1.4 Tariff support

The lack of long-term or stable policy commitments from governments is another significant barrier as it affects developer and market confidence. Furthermore, in many countries, the level of financial support provided (feed-in tariff or tradable certificates) often appears either insufficient or, at best, marginal in order to provide sufficiently attractive returns to investors compared to lower risk investment options in other sectors.

Confidence in long-term market opportunities is required from the private sector in order to trigger the investment decisions necessary for the development of a supply chain for the offshore renewable sector. At the national level, governments can heavily influence and coordinate the development of required infrastructures, such as harbours and grid. The importance of public support for marine technologies is illustrated by the success that offshore wind has had in some countries. Belgium, Denmark, Germany and the UK, where governments have shown strong political leadership and tailored financial incentives, are leading the way in terms of deployment.

11.1.5 Planning and permitting

Complex permitting processes are another major barrier to offshore renewable energy project development in most countries. Prescriptive planning conditions or requirements limit project and technology design options and can significantly increase timescale and development costs. A number of regulatory barriers are also delaying or preventing the changes required in the onshore and offshore grid infrastructure in order to accommodate offshore (and onshore) renewable expansion plans. Deployment timescales can be greatly increased as a result of these barriers.

While the permitting processes are diverse and country specific, lessons can be learned from the countries that have had more success with offshore wind. Streamlined application procedures, one-stop shops and pre-permitted areas are some of the potential mitigation measures to planning and permitting barriers. The allocation of seabed rights to competent and construction-focused developers is also important in order to prevent sites from being leased to developers more interested in speculative applications or without the necessary resources to progress the development of projects.

11.1.6 Technology

The main technical challenges and barriers shared by all marine renewable energy technologies include technology and design optimization, reliability, installation and decommissioning, operation and maintenance, grid connection and integration. Considerable investments will be required in onshore and offshore grid infrastructure in order to accommodate the large expected expansion in variable generation capacity from offshore renewable energy projects. The optimal topology of this expansion needs to be considered at a continental rather than national or Eurocentric level. The treatment of offshore networks on a supranational level should be aimed at in the longer term. Technical barriers are surmountable but usually affect the cost of offshore renewable energy project and technologies.

11.1.7 Research, development, demonstration and green employment

Research, development and demonstration (RD&D) activities performed directly by the private sector or financially supported or promoted by public funding are instrumental to the removal or mitigation of technical barriers; by creating domestic intellectual capital, they can also support green employment and the development of future industries. The importance of the support that can be provided by publicly funded RD&D activities is particularly relevant for the more immature technologies given the lower investment capacity of the private sector and longer timescales involved. Direct involvement and possibly co-investment from private companies into RD&D activities should be maximized.

11.1.8 Other barriers

Other barriers include health-and-safety, environmental and other sea-user considerations, supply chain constraints and skills shortages. While all of these issues are important and have to be dealt with, mitigation measures can reduce their impact.

The reduction or mitigation of health-and-safety barriers can be achieved by a strong industry culture, supported by staff training, compliance with legislative requirements, best practices and standards, as well as through technical innovations.

The main mitigation measures in order to ensure the minimization of environmental barriers and acceptance from other sea users are early engagement with all stakeholders, appropriate marine spatial planning and adoption of the recommendations from the project-specific environmental impact assessment (EIA).

The removal of skills barriers requires the active promotion of the various employment and careers opportunities provided by the offshore renewable energy industry, as well as the development of training courses and programmes tailored to the needs of the industry.

11.2 Project development recommendations

Developers are recommended to follow best practice at all stages of a project's lifecycle, from pre-feasibility to development, design, construction, operation and decommissioning.

Such best practices are included in the main body of this book and detail the steps required to minimize risks associated with poor development (safety, environmental, delays and costs).

11.3 Model policy framework

The development and application of a clear policy framework can play a vital role in securing investment in offshore development and, over time, in helping to support innovation, a competitive environment and to reduce costs.

From the developer's perspective, a supportive national offshore renewable deployment strategy would offer:

- a strategic policy framework;
- measures that foster innovation and competition;
- an efficient permitting system;
- environmentally and socially sustainable development;
- grid connection;
- favourable power market; and
- commercially viable investment opportunities.

These items are discussed in more detail in the following sections. A larger number of references are available regarding renewable energy policies in general. This model policy framework and its components are based on elements and parts of policies and programmes that have proved to be successful in various countries for offshore renewable energy technologies or in similar conditions (new, less mature energy technologies in emerging markets) if experiences in offshore are not yet available. The model is also based upon best practices and knowledge available to date and upon technical, economic and policy conditions that have proved to be conducive to the deployment of new technologies. Assessing the effectiveness of implemented policies should also be of prime importance in order to quickly address any identified shortcoming or adapt to changing conditions. Relevant examples for further reference can be found in the OECD/IEA (2008) and REN 21 (2010) global status reports.

The replication potential of the model policy framework proposed hereafter will vary from country to country as a function of:

- the stage of existing or on-going policy development related to offshore renewable energy technologies (from very advanced, as in the UK, to non-existent) or other offshore industries;
- status and potential for the deployment of offshore renewable technologies;
- existence or absence of additional economic benefits (such as supply chain opportunities) associated with the deployment of offshore renewable energy technologies.

Depending upon these factors, the replication potential in a given country can be very high ('blank sheet') or seriously limited, and needs to be assessed by policy-makers on a case-by-case basis.

11.3.1 Strategic policy framework

Offshore renewable technologies can play an important role in meeting policy objectives of cleaner and more secure sources of energy combined with economic opportunity. The pursuit and achievement of these objectives has to be primarily balanced against the high investment costs currently involved with offshore renewable energy technologies. Each region faces unique circumstances that shape the scope for an offshore development programme. For example, UK policy-makers have pushed offshore development due (amongst others) to limited onshore options and the availability of large potential zones favourable to marine energy development. Norway, on the other hand, is facing a reverse set of circumstances, with abundant onshore development options and deeper seabed; consequently, Norway appears to be focusing efforts on longer-term innovation in floating wind turbine technologies.

A model overarching policy framework would consist of:

- *long-term and consistent policy*, translated into regulation, essential in order to underpin investor confidence for projects that will span decades;
- *a combined set of regulatory drivers* based on greenhouse gas (GHG) reductions, energy security and technology-specific deployment measures;
- *compulsory GHG reduction targets* – these, by establishing short- and long-term limits on GHG emissions and a price on carbon, play an important role in driving strategic investment by government and utility companies; internationally binding targets (e.g. EU and post-Kyoto agreement) reinforce national government obligations to decrease emissions;
- *energy security* policy and regulation, in the form of renewable energy and grid development measures complement GHG reduction targets and play an important role in the development of the next generation of transmission networks, including breaking down regional technical and market constraints;
- *technology deployment* policy is required to steer investment and activities towards strategic areas such as offshore renewables and grid network enhancements, and to prevent GHG reduction targets from being achieved purely through short-term measures such as energy efficiency and carbon trading.

A model policy framework would, for example, combine a set of regulatory drivers in order to reduce opportunities for rapid policy changes. For example, the EU '20-20-20' target (European Union Climate Action, 2010) commitments provide consistent pressure on member states to achieve GHG and renewable energy targets.

A model policy framework would also reconcile regional versus national policy on key issues such as planning policy and development and management of grid networks and power markets. This is a key issue which is particularly important to the successful implementation of deployment programmes in federal countries.

The underpinning financial policy framework also needs to be robust, workable and sustainable over the long term. This includes the nation's credit rating, the design of fiscal support mechanisms and allocation of grid development costs.

11.3.2 Innovation and competition

A strategic policy approach to innovation is important in order to support developing wave and tidal technologies, as well as further developments in offshore wind energy.

A model policy framework would address the critical stages of commercialization, such as:

- research, development and demonstration;
- commercial scale-up;
- market entry; and
- market diffusion.

Specific elements are outlined below.

Research, development and demonstration policy initiatives can include:

- government grants;
- tax incentives;
- incubator programmes;
- fostering collaboration between research, government and commercial bodies;
- testing and certification regimes, including test facilities, strategic and financial support, and flexible verification regimes to provide appropriate purpose testing.

Commercial scale-up and market-entry policy initiatives include:

- capacity-building: business and intellectual property advice and technology transfer;
- low-interest loans;
- loan guarantees;
- grants;
- technology clusters, innovation parks and other infrastructure support;
- tax incentives or regulatory requirements for adoption of innovative technology;

- public procurement mechanisms;
- financial support for certification; and
- avoidance of restrictive standards and regulations.

An effective regime would foster the scaling-up of businesses to reduce costs and risk, while still fostering competition and access for new entrants and technological innovation in the market. The regime would also support the production of local jobs, skills and industry without undermining sector competitiveness and development on a global scale, which is ultimately required to harmonize approaches, build economy of scale and reduce costs.

Several measures to help achieve this objective include:

- develop and apply internationally adopted standards, as long as these are not prescriptive to the point that they block out innovation or competition;
- establish alternative validation regimes to enable new technologies and innovations to demonstrate performance to the extent required without the need to fully comply with standards applicable to commercial technologies;
- provide incentives to large businesses for the adoption of innovative technologies or services from small or local businesses;
- avoid local content rules (LCRs) (see WRI, 2009) and consider less prescriptive means to support the local economy (the WRI reports that job creation would be created anyway and global integration is required to reduce costs; WRI, 2009);
- avoid fiscal support instruments that offer an advantage to local businesses;
- award tenders to multiple businesses in order to foster competition as businesses scale up.

An offshore energy regime will only be effective if it diffuses the market to a point that a meaningful reduction in fossil fuel-based power is achieved. This requires the establishment of a pipeline of projects large enough to encourage commercial interest and supply chain investment.

11.3.3 Permitting efficiency

An effective permitting regime requires the following two key elements:

- consistent and supportive policy across government decision-makers; and
- a clear, transparent and efficient process.

An effective permitting framework would pre-empt potential policy inconsistencies and inefficiencies across national, regional or federal government departments. Specific measures would include:

- unified offshore policy and priorities across government departments;
- mechanisms that incentivize government departments, including regional authorities to adopt offshore renewable energy;
- national planning policy and authority;
- national electrical transmission policy and authority;
- unified maritime permitting authority;
- appropriate regulation that addresses requirements of offshore energy generation.

A model permitting framework would include:

- government requirements to streamline permitting procedures;
- the presence of a regulatory body with the remit and authority to address permitting inefficiencies (e.g. UK Cabinet Office Better Regulation Task Force);
- staged permitting process to reduce the risk and upfront design work required associated with permitting;
- establishment of preapproved sites to reduce permitting risk, attract construction-oriented developers and provide a signal to investors on the scale of offshore development being seriously pursued;
- trialled and documented permitting procedures;
- single or coordinating permitting agency ('one-stop shopping approach');
- reduced permitting requirements for demonstration projects, especially avoiding the need to re-permit installations, following minor design changes;
- adequate regulator resources, including expertise;
- stakeholder and regulator support structure to facilitate an understanding of key processes and issues;
- application of harmonized approaches, where helpful.

11.3.4 Environmentally and socially responsible development

Establishing that offshore energy deployment has an acceptable impact upon the environment and society is essential for gaining planning permissions. A model framework would:

- take into account the positive environmental and social impacts of proposed developments in EIAs, based on the whole lifecycle impacts of offshore generation versus conventional sources;
- provide robust maritime data on baseline conditions and impacts of developments to inform risk-based decisions, without the unnecessary application of the precautionary principle;

- encourage public acceptance of renewable energy, through education programmes, and other regimes, such as localized small-scale developments under feed-in tariff regimes;
- allow opportunity to compensate for local impacts by enhancing environmental conditions elsewhere, as a last resort.

11.3.5 Grid connection

The model framework would offer prompt connection to the onshore grid network for offshore renewable energy with the following proposed characteristics.

Grid development

- National grid development and planning policy, regulation and authority which has overcome regional policy impediments and transmission capacity constraints.
- Prioritization of grid development projects to support renewable energy through measures such as:
  - complete unbundling of generation and transmission entities;
  - mechanism to prevent applications for questionable projects from holding up grid development for renewable projects;
  - regulation.
- Streamlined permitting mechanism.
- National allocation of deep connection costs.

Mechanisms to facilitate grid connections

- Loan guarantees to break the deadlock between transmission system operators (TSOs) and offshore developers.
- Passing constraint costs on to rate payers.
- Technical options, such as dynamic ratings and direct current (DC) connections.

11.3.6 Favourable power market

The model framework would offer favourable market conditions for offshore renewable energy with the following proposed characteristics:

- short-term wind generation forecasting;
- short-term time between gate closure and real-time trading;
- availability of aggregator (which pools risks across other parties); and
- long-term feed-in tariffs to compensate for a lack of long-term power purchase agreements (PPAs).

11.3.7 Commercially viable investment opportunity

A model offshore renewable energy reward/risk ratio would be high enough to attract investment by the parties required for project development, including project developers, lenders, owners such as utility companies, and supply chain members.

A model framework would also include stable and sufficient fiscal support instruments, such as feed-in tariffs or tradable certificates. Higher rates may need to be provided from more risky or costly projects (e.g. deep water or distant developments). An effective regime would combine the advantages of feed-in tariffs with tradable certificates.

Fiscal incentives that reward renewable energy generators would be complemented with downside fiscal drivers, such as compulsory allowance or tradable certificate purchases or penalties.

Government or multilateral loan guarantees are also currently a critical element for gaining debt finance.

Project risks would be mitigated by elements outlined in this book, especially:

- a strong government credit rating;
- a clear, transparent and workable permitting regime;
- pre-selection and assessment of sites;
- a tendering structure that reduces contract interface risk;
- grid connection certainty and priority access;
- fair competition; and
- cost-control measures.

11.3.8 Policy framework conclusions

Offshore renewable technologies can play an important role in meeting the policy objectives of cleaner and more secure sources of energy combined with economic opportunity. Each country or region faces unique circumstances that shape their decision to support or not the development and deployment of offshore renewable energy technologies and projects, as well as the scope of an offshore development programme.

A policy framework should receive strong visible support from government and government organizations in order to emphasize the commitment to the industry. Nevertheless, once the projects are in the construction phase, government intervention should be kept to the minimum.

Appendix A
References

ADEME (Agence de l'Environnement et de la Maitrise de l'Energie) (2009) *Appel à Manifestations d'interet (AMI): Demonstrateurs de Recherche en Nouvelles Technologies de l'énergie (NTE) – Energies marines*, http://www2.ademe.fr/servlet/getDoc?cid=96&m=3&id=62661&p1=1

AEA/SEI (2006) *Review and Analysis of Ocean Energy Systems Development and Supporting Policies*, IEA-Oceans, http://www.iea-oceans.org/_fich/6/Review_Policies_on_OES_2.pdf

Aqua Buoy Finevara Renewables Ltd (2010) http://www.celsias.com/media/uploads/admin/aquabouy.jpg

Aquatera Ltd (2008) *Announcement (8 December 2008): Aquatera and ITRI Sign Contract to Conduct an Evaluation Study into the Development of Wave Energy in Taiwan*, http://www.aquatera.co.uk/news_detail.asp?ID_News = 1

Archer, C. L. and Jacobson, M. Z. (2005) 'Evaluation of global wind power, Stanford University', *Journal of Geophysical Research*, vol 110, http://www.stanford.edu/group/efmh/winds/2004jd005462.pdf

Arup Energy/The Carbon Trust (2005) *Marine Energy Challenge: Oscillating Water Column Wave Energy Converter Evaluation Report*, Carbon Trust, http://www.carbontrust.co.uk/SiteCollectionDocuments/Various/Emerging%20technologies/Technology%20Directory/Marine/Other%20topics/OWC%20report.pdf

AWEA (American Wind Energy Association) (2008) *Wind Energy for a New ERA: An Agenda for the New President and Congress*, American Wind Energy Association, http://www.newwindagenda.org/documents/Wind_Agenda_Report.pdf

AWEA/Solar Energy Industries Association (2009) *Green Power Superhighways: Building a Path to America's Clean Energy Future*, American Wind Energy Association and the Solar Energy Industries Association, http://www.hks.harvard.edu/hepg/Papers/2009/GreenPowerSuperhighways_0209.pdf

Barr, P. (2009) 'Overview of conference proceedings', UK–Taiwan Workshop on Tidal Current Energy, Royal Society of Edinburgh, Edinburgh, http://www.ma.hw.ac.uk/RSE/events/reports/2008-2009/taiwan.pdf

Beaudoin, G., Robertson, D., Doherty, R., Corren, D., Staby, B. and Meyer, L. (2010) 'Technological challenges to commercial scale application of marine renewables', *Oceanography*, vol 23, no 2, Special Issue on Marine Renewables, http://www.tos.org/oceanography/issues/issue_archive/issue_pdfs/23_2/23-2_beaudoin.pdf

BERR (Department for Business Enterprise and Regulatory Reform) (2008) *Atlas of UK Marine Renewable Energy Resource: A Strategic Environmental Assessment Report*, ABPmer, The Met Office and Proudman Oceanographic Laboratory, http://www.renewables-atlas.info/downloads/documents/Renewable_Atlas_Pages_A4_April08.pdf

BERR (2009a) *UK Ports for the Offshore Wind Industry: Time to Act* (URN 09/511X), http://www.berr.gov.uk/files/file49871.pdf

BERR (2009b) *UK Offshore Wind Ports Prospectus* (URN 09D/562), http://www.berr.gov.uk/files/file51407.pdf

Besio, G. and Losada, M. A. (2008) 'Sediment transport patterns at Trafalgar offshore windfarm', *Ocean Engineering*, vol 35, pp653–665

Biehl, F. and Lehmann, E. (2006) 'Collision risk analysis and collision friendly design of offshore wind farms', in *Proceedings of OMAE 2006, Hamburg: Collisions of Ships and Offshore Wind Turbines – Calculation and Risk Evaluation*, http://www.tu-harburg.de/t3resources/skf/Dokumente/Lehmann-Biehl-OMAE2006.pdf

Black & Veatch/The Carbon Trust (2005) *Marine Energy Challenge: Phase II, UK Tidal Stream Energy Resource Assessment*, Carbon Trust, http://www.carbontrust.co.uk/SiteCollectionDocuments/Various/Emerging%20technologies/Technology%20Directory/Marine/Other%20topics/PhaseIITidalStreamResourceReport.pdf

Blue H Group (2010) www.bluehgroup.com/index.php

Blue Water (2010a) http://www.bluewater.com/products.asp

Blue Water (2010b) *CALM Buoy for the Western Libyan Gas Project Wafa Plant*, Blue Water, http://www.bluewater-offshore.com/downloads/CALM%20Buoy_WLGP.pdf

Boehlert, G. W. and Gill, A. B. (2010) 'Environmental and ecological effects of ocean renewable energy development: A current synthesis', *Oceanography*, vol 23, no 2, pp68–81, http://www.tos.org/oceanography/issues/issue_archive/issue_pdfs/23_2/23-2_boehlert.pdf

Broughton, P. (2010) 'Decommissioning of the Maureen Oil Platform', *Ingenia*, http://www.ingenia.org.uk/ingenia/issues/issue6/Broughton.pdf

Brower, M., Bailey, B. and Zack, J. and AWS Truewind (2006) *New High-Resolution Wind Resource Maps of China*, http://www.awstruepower.com/wp-content/media/2010/05/WP_Asia_2006_Mapping2.pdf2.pdf

BSH (2007) 'Investigation of the impacts of offshore wind turbines on the marine environment' (StUK 3), p9, http://www.bsh.de/en/Products/Books/Standard/7003eng.pdf, accessed 21 September 2010

BSH (2010) *Offshore Wind Standards*, http://www.bsh.de/en/Products/Books/Standard/index.jsp, accessed 22 September 2010

BTM Consult Aps (2010) *Supply Chain Assessment 2010*, http://www.btm.dk/reports/supply+chain+assessment+2010

BTM Consult Aps (2011) *International Wind Energy Development: World Market Update 2010*, March

BWE (Bundesverband WindEnergie e.V) (2010) *Offshore Wind Potential*, http://www.wind-energie.de/de/technik/entstehung/windpotential/, accessed 12 February 2010

BWEA (British Wind Energy Association) (2008a) 'Employment opportunities and challenges in the context of rapid industry growth', Figure 6, p10, http://www.bwea.com/pdf/publications/Bain%20Brief_Wind%20Energy%202008_FINAL.pdf

BWEA (2008b) 'Today's investment – tomorrow's asset: Skills and employment in the wind, wave and tidal sectors', Figure 2.1, http://www.bwea.com/pdf/publications/BWEA%20Skills%20Report%20FINAL%2016oct.pdf

BWEA/Entec (2009) *Marine Renewable Energy: State of the Industry Report*, BWEA, http://www.bwea.com/pdf/marine/Marine_report_enteclogo.pdf, October 2009

Canadian Hydraulics Centre (2006) *Inventory of Canada's Marine Renewable Energy Resources*, Canadian Hydraulics Centre, http://www.oreg.ca/web_documents/chc-tr-041.pdf

Carballo, R., Iglesias, G. and Castro, A. (2008) 'Numerical model study of tidal energy potential in the Ria de Muros', in *Proceedings of ICCE 2008*, Hamburg, 31 August–5 September 2008

Carbon Trust (2006) Future of Marine Energy: Results of the Marine Energy Challenge – Cost Competitiveness and Growth of Wave and Tidal Stream Energy, http://www.oceanrenewable.com/wp-content/uploads/2007/03/futuremarineenergy.pdf

Carbon Trust (2010) *Offshore Wind Accelerator Foundation Programme*, http://www.carbontrust.co.uk/emerging-technologies/current-focus-areas/offshore-wind/pages/owa-competition1.aspx, accessed 7 September 2010

CIM (Cambridge Institute of Manufacturing) (2006) *University of Cambridge, Analysis of Environmental Firms*, N. Dee and S. Ford, Cambridge

CMA (China Meteorological Association) and Sgurr Energy (forthcoming) *Offshore Wind Energy Resource Assessment and Feasibility Study of Offshore Wind Farm Development in China*, EuropeAid/123310/D/SER/CN

Chuang, R., Schafer, J. and Speht, R. (2009) *Opportunities and Challenges of Kuroshio Current Energy Development in Taiwan*, Internal report

CIA (2010) *The World Fact Book*, 'World economy', https://www.cia.gov/library/publications/the-world-factbook/geos/xx.html, accessed 18 March 2010

CleanTech (2010) http://cleantech.com/news/4276/pelamis-sinks-portugal-wave-power-p

Clément, A., McCullen, P., Falcão, A. et al (2002) 'Wave energy in Europe: Current status and perspectives', in *Renewable and Sustainable Energy Reviews*, Elsevier, vol 6, pp405–431, http://www.elsevier.com/wps/find/journaldescription.cws_home/600126/description#description

CCA (Coastal Concern Alliance) (2010) 'Need for reform: Offshore windfarms', http://www.coastalconcern.ie/need_for_reform/offshore_windfarms.html

COWRIE (Collaborative Offshore Wind Research into the Environment) (2006) Effects of Offshore Wind Farm Noise on Marine Mammals and Fish (F. Thomsen, K. Luedemann, R. Kafemann and W. Piper), http://www.offshorewindfarms.co.uk/Pages/COWRIE

COWRIE (2010) http://www.offshorewindfarms.co.uk, accessed 24 August 2010

Couch, S. J. (2010) 'Tidal current energy: Improved methodologies for resource assessment', Presentation at SuperGen Marine Energy '10, 17 November, 2010, www.supergen-marine.org.uk/drupal/content/past-events/assembly-2010

Dalton, G., Rousseau, N., Neumann, F. and Holmes, B. (2009) 'Non-technical barriers to wave energy development, comparing progress in Ireland and Europe', in *Proceedings of the 8th European Wave and Tidal Energy Conference*, Uppsala, Sweden

Danish Energy Agency (2010) http://www.ens.dk/en-US/supply/Renewable-energy/WindPower/offshore-Wind-Power/Sider/Forside.aspx, accessed 10 February 2010

Danish Energy Authority (2007) *Future Offshore Wind Power Sites – 2025*, Committee for Future Offshore Wind Power Sites, http://193.88.185.141/Graphics/Publikationer/Havvindmoeller/Fremtidens_%20havvindm_UKsummery_aug07.pdf

Degraer, S. and Brabant, R. (eds) (2009) *Offshore Wind Farms in the Belgian Part of the North Sea: State of the Art after Two Years of Environmental Monitoring*, Royal Belgian Institute for Natural Sciences, Management Unit of the North Sea Mathematical Models, Marine Ecosystem Management Unit, Belgium

DECC (Department of Energy and Climate Change) (2009) *Cost of and Financial Support for Offshore Wind*, Ernst and Young report, URN 09D/534, www.bis.gov.uk/files/file51142.pdf

DECC (2010a) *Proposals for Improving Grid Access Impact Assessment*, 27 July 2010, http://www.decc.gov.uk/assets/decc/consultations/improving%20grid%20access/253-improving-grid-access-ia.pdf

DECC (2010b) *Renewables Obligation*, www.decc.gov.uk/en/content/cms/what_we_do/uk_supply/energy_mix/renewable/policy/renew_obs/renew_obs.aspx, accessed 12 February 2010

DECC (2010c) *UK Electricity Generation Costs Update*, Mott MacDonald report, http://www.decc.gov.uk/en/content/cms/statistics/projections/projections.aspx

Den Boon, H., Just, H., Friis Hansen, P., Sonne Ravn, E., Frouws, K., Otto, S., Dalhoff P., Stein, J., van der Tak, C. and van Rooij, J. (2004) 'SAFESHIP: Reduction of ship collision risks for offshore wind farms', in *Conference Proceedings to the European Wind Energy Conference (EWEC) 2004*, http://www.2004ewec.info/files/24_1600_henkdenboon_01.pdf

Department of the Environment, Heritage and Local Government (2010) *Foreshore*, www.environ.ie/en/Foreshore/, accessed 9 February 2010

Deutsche Windguard (2010) *Draft Report on Task1: Accelerating the Deployment of Offshore Renewable Energy Technologies (ADORET)*, DeutscheWindguard

DGEC (Ministère de l'Écologie, de l'Énergie, du Développement durable, et de la Mer) (2010) http://www.developpement-durable.gouv.fr/energie/renou/f1e_en.htm, accessed 10 February 2010

DOE (US Department of Energy) (2009) *Wave Dragon: Marine and Hydrokinetic Technology Database*, www.eere.energy.gov/windandhydro/hydrokinetic

Douglas Westwood (2007) *POWER (Pushing Offshore Wind Energy Regions)*, http://www.offshore-power.net/Files/Dok/2007-06%20power%20-%20transnational%20study%20update.pdf

DTI (Department of Trade and Industry) (2002) *Future Offshore: A Strategic Framework for the Offshore Wind Industry*, DTI, UK

DTI (2004a) *Floating Attenuator Wave Energy Device, Wavegen Hydra Project*, URN No 04/1087, DTI, UK

DTI (2004b) *Renewable Supply Chain Gap Analysis*, DTI, UK

DTI (2007) *Impact of Banding the Renewable Obligations: Cost of Electricity Production*, URN 07/948. http://www.berr.gov.uk/files/file39038.pdf

Dudziak, G. (2009) 'Comparison of emerging renewable energy technologies', C5 Renewable Energy Investments and Financing Forum, London

Duwind (Delft University Wind Energy Research Institute) (2001) *Offshore Wind Energy: Ready to Power a Sustainable Europe*, Delft, the Netherlands

EC (European Commission) (2005) *Concerted Action for Offshore Wind Energy Deployment (COD), Principal Findings 2003–2005*, EC, Brussels

EEA (European Environment Agency) (2009) *Europe's Onshore and Offshore Wind Energy Potential: An Assessment of Environmental and Economic Constraints*, EEA, Brussels

EIA (2010) *RENBAR: Good Practices for Solving Environmental, Administrative and Socio-Economic Barriers in the Deployment of Renewable Energy Systems*, IEA–REETD, http://www.iea-retd.org/files/ToR_RENBAR_090710.pdf

Enova (2007) *Potensialstudie av havenergi i Norge* [*Resource Potential for Ocean Energy Production in Norway*], http://www.enova.no/PUBLIKASJONSOVERSIKT/publicationdetails.aspx?publicationID=266

Enova (2010) www.enova.no, accessed 17 February 2010

ENS (2010) Energistyrelsen (Danish Energy Agency) Environmental Monitoring Programme, http://www.ens.dk/en-US/supply/Renewable-energy/WindPower/offshore-Wind-Power/Environmental-impacts/Sider/Forside.aspx

Ente Vasco de la Energia (2009) 'Actions in the Basque Country (Spain) in marine energy', Presentation at the National Maritime Museum, Falmouth, Cornwall, 22–23 October 2009

Environment Canada (2010) *Canadian Wind Energy Atlas*, http://www.windatlas.ca/en/maps.php, accessed 10 February 2010

Environmental Science Division (2010) 'Case history: Soil investigation for offshore suction anchors', http://www.ead.anl.gov/new/newsdocs/Lunne_lect6_suction_anchor.pdf

EPRI (Electric Power Research Institute) (2009) *Testimony: Marine and Hydrokinetic Technologies: Finding the Pathway to Commercialisation*, Roger Bedard, http://gop.science.house.gov/Media/hearings/energy09/dec3/Bedard.pdf

EPRI and Bedard, R. (2007) *Overview of Wave and Current Energy: Resource, Technology, Environmental and Business Issues*, Louisiana State University (LSU), Center for Energy Studies, Conference, 25 April 2007, http://www.enrg.lsu.edu/Conferences/altenergy2007/bedard.pdf

EPRI and Bedard, R. (2009) 'Presentation on US ocean or marine energy resources and potential', http://www.stanford.edu/~mpinsky/sjlsp/Bedard%20-%20Marine%20Renewables.pdf

EREC (European Renewable Energy Council) (2009) *RES National Policy Reviews*, http://www.erec.org/policy-actions/national-policies.html

EREC (2010) *Emission Trading Scheme and NER300*, http://www.erec.org/index.php?id=310, accessed 20 October 2010

ETI (Energy Technology Institute) (2010) http://www.energytechnologies.co.uk/Home/Partners/Current-Members.aspx, accessed 20 October 2010

ETSAP (2010) *Technology Brief E02 Gas-Fired Power*, http://www.etsap.org/E-techDS/PDF/E02-gas_fired_power-GS-AD-gct.pdf

EurObserv'ER (2010) *The State of Renewable Energies in Europe 2009: 9th EurObserv'ER Report*, http://www.eurobserv-er.org/pdf/barobilan9.pdf

European Union Climate Action (2010) http://ec.europa.eu/clima/policies/eu/package_en.htm, accessed 21 October 2010

EWEA (European Wind Energy Association) (2009a) *The Economics of Wind Energy*, EWEA, http://www.ewea.org/index.php?id=201

EWEA (2009b) *Ocean of Opportunity: Harnessing Europe's Largest Domestic Energy Source*, September 2009

EWEA (2010a) *The European Offshore Wind Industry: Key Trends and Statistics 2009*, EWEA, January, http://www.ewea.org/fileadmin/emag/statistics/2009offshore/pdf/offshore%20stats%2020092.pdf

EWEA (2010b) *Wind Direction September 2010*, p46, http://www.ewea.org/fileadmin/emag/winddirections/2010-09/pdf/100813_WD%20Sep_FINAL.pdf

E&Y (Ernst and Young) (2010) *Renewable Energy Country Attractiveness Indices*, issue 25

Farber Maunsell/Scottish Executive (2007a) *Scottish Marine Renewables SEA: Environmental Report Section B: Marine Renewables – Resource and Technology*, http://www.seaenergyscotland.net/public_docs/ER_SectionB_Final.pdf (for information links to section A, see http://www.seaenergyscotland.net/public_docs/ER_SECTION_A_&_FrontCover_MAR07.pdf to section C, see http://www.seaenergyscotland.net/public_docs/ER_C19_Landscape_Seascape_&Appendicies.pdf)

Farber Maunsell/Scottish Executive (2007b) *Offshore Wind Power Deployment in Germany*, http://erneuerbare-energien.de/files/pdfs/allgemein/application/pdf/offshore_wind_deployment_de_en.pdf

Farber Maunsell/Scottish Executive (2009) *Tariffs and Sample Degression Rates Pursuant to the New Renewable Energy Sources Act (Erneuerbare-Energien-Gesetz – EEG) of 25 October 2008*, http://erneuerbare-energien.de/files/english/pdf/application/pdf/eeg_2009_verguetungsdegression_en_bf.pdf

First Subsea (2010) *Ballgrab Subsea Mooring Systems*, http://www.firstsubsea.com/fspages/mooring/subsea-mooring-systems.htm?case_study

Fisheries and Oceans Canada (2010) www.dfo-mpo.gc.ca/oceans/industries/oceanenergy-energiemarine/renewable-renouvelable-eng.htm, accessed 16 February 2010

FREDS Marine Energy Group (2009) *Marine Energy Group Supply Chain Survey, 2009*, Sgurr Energy and IPA Energy and Economics report, http://www.cambridge-resource-economics.co.uk/MEG%20Supply%20Chain%20Survey%20Report.pdf; FREDS presentation: http://www.scotland.gov.uk/Resource/Doc/281865/0085187.pdf

Fri-El Green Power (2010) http://www.fri-el.it/seapower/en/project_sea_power_device.php, accessed 11 February 2010

Fugro Oceanor (2008) *World Waves Atlas*, www.oceanor.com

German Environment Ministry (Federal Ministry for the Environment, Nature Conservation and Nuclear Safety) (2010) 'RES LEGAL: Legislation on renewable energy generation', http://res-legal.eu/en.html

GL (2002) *Richtlinie zur Erstellung von technischen Risikoanalysen für Offshore-Windparks*, GL

Guillet, J. (2009) *Financing Offshore Wind Farms: What Can Be Done and What Works*, C5 Renewable Energy Investments and Financing Forum, London

GWEC (Global Wind Energy Council) (2010) www.gwec.net, accessed 10 February 2010

Hammerfest-strøm (2010) http://www.hammerfeststrom.com/content/view/45/72/lang,en/

Harris, R. E., Johanning, L. and Wolfram, J. D. (2004) 'Mooring systems for wave energy converters: A review of design issues and choices', in *Proceedings of the 3rd International Conference on Marine Renewable Energy (MAREC)*, Blyth, UK, 2004

HM Government (2009) *The UK Renewable Energy Strategy*, Her Majesty's Stationary Office, London

HSE (Health and Safety Executive) (2010) *Emerging Energy Technologies Programme (EET)*, http://www.hse.gov.uk/eet/about.htm, accessed 21 September 2010

IEA (International Energy Agency) (2006) *Renewable Energy: RD&D Programmes – Insights from IEA Technology Programmes*, http://www.iea.org/textbase/nppdf/free/2006/renewenergy.pdf

IEA (2009a) Technology Roadmap: Wind Energy, IEA, http://www.iea.org/Papers/2009/wind_roadmap.pdf

IEA (2009b) *Electricity Information 2009*, IEA (see http://www.planbleu.org/portail_doc/electricity_information2010.pdf for the more recent 2010 report)

IEA (2010) *Policies and Measures Database*, http://www.iea.org/textbase/pm/index.html, accessed 20 October 2010

IEA–OES (International Energy Agency–Ocean Energy Systems) (2009a) *Annual Report 2008*, http://www.iea-oceans.org/_fich/6/Annual_Report_2008_(1).pdf

IEA–OES (2009b) *Wave Energy Pilot Zone,* Portugal, Teresa Pontes, LNEG, National Activities, IEA OES Newsletter October 2009, issue 12

IEA Wind (International Energy Agency Wind) (2009) *IEA Wind Annual Report*, http://www.ieawind.org/AnnualReports_PDF/2009.html

IEA Wind (2010) *Task 23: Offshore Wind Energy Deployment*, www.ieawind.org/Annex_XXIII.html, accessed 16 February 2010

IMCA (International Marine Contractors Association) (2010) *Basic Principles of Dynamic Positioning*, http://www.imca-int.com/divisions/marine/reference/intro02.html

IMO (International Maritime Organization) (2010) *Status of Conventions by Countries*, http://www.imo.org/includes/blastDataOnly.asp/data_id%3D29527/status-x.xls, accessed 22 September 2010

International Water Power and Dam Construction (2010) 'Spain launches marine energy research initiative', http://www.waterpowermagazine.com/story.asp?sectionCode=130&storyCode=2055205

Invest in Canada (2010) http://investincanada.gc.ca/eng/industry-sectors/renewableenergy.aspx, accessed 9 February 2010

JAMSTEC (Japan Agency for Marine-Earth Science and Technology) (2010) www.jamstec.go.jp/e/, accessed 11 February 2010

Kjeller Vindteknikk (2009) *Vindkart for Norge. Kartbok 1a. Årsmiddelvind i 80 m høyde*, http://www.nve.no/no/energi1/fornybar-energi/vindkraft/

Lane, N. (2008) *Congressional Research Service (CRS) Report for Congress: Wave, Tidal and In-Stream Energy Projects: Which Federal Agency Has the Lead?*, http://assets.opencrs.com/rpts/RS22721_20081126.pdf

Leonardo Energy (2010) *Blue Energy*, http://www.leonardo-energy.org/drupal/book/export/html/2243, accessed 16 February 2010

Lockheed Martin (2010) *Ocean Thermal Energy Conversion (OTEC)*, http://www.lockheedmartin.com/products/OTEC/index.html, accessed 16 February 2010

LMC (London Marine Consultants) (2010) *CPTL 137B 'Farwah' FPSO*, http://www.londonmarine.co.uk/resources/file/lmc_cptl.pdf

Lyard, F., Lefevre, F., Letellier, T. and Francis, O. (2006) 'Modelling the global ocean tides: Modern insights from FES2004', *Ocean Dynamics*, no 56, pp394–415, http://maia.usno.navy.mil/conv2010/chapter6/add_info/FES2004.pdf

Mankins, J. C. (1995) *Technology Readiness Levels: A White Paper*, NASA, Office of Space Access and Technology, Advanced Concepts Office, http://www.hq.nasa.gov/office/codeq/trl/trl.pdf

Marimatech (2010) *Offshore*, http://www.marimatech.com/?show=offshore

Marine Institute, Sustainable Energy Ireland (2002) *Options for the Development of Wave Energy in Ireland*, Public Consultation, Ireland

Marine Institute, Sustainable Energy Ireland (2005) *Accessible Wave Energy Resource Atlas Ireland, Revision 4*, Ireland

Mehta, R. (2009) 'Survival of the fittest in wave energy', *Materials World Magazine*, 1 June 2009

METI (Ministry of Economy, Trade and Industry) (2010) www.meti.go.jp/english/policy/energy_environment/renewable/index.html, accessed 15 February 2010

Milborrow, D. (2010) 'Annual power cost comparison', *Wind Power Monthly*, http://www.windpowermonthly.com/news/975661/Annual-Power-Costs-Comparison-difference-year, accessed 13 July 2010

Ministry of the Environment and Ministry of Enterprise, Energy and Communications (2009) *An Integrated Climate and Energy Policy*, www.sweden.gov.se/content/1/c6/12/34/66/1a1aa683.pdf

Natural Resources Canada (2010) http://www.nrcan.gc.ca/eneene/science/renren-eng.php, accessed 9 February 2010

NCOR (National Center for Ocean Research) (2010) *Ocean Data Bank, Current Velocity Data 1991 to 2008*, http://140.112.65.17/odbs/Eng/index-e.html, accessed 17 February 2010

NDRC (National Development and Reform Commission) (2005) *The Renewable Energy Law of the People's Republic of China*, http://www.npc.gov.cn/englishnpc/Special/CombatingClimateChange/2009-08/25/content_1515301.htm

NDRC (2007) *Medium- and Long-Term Development Plan for Renewable Energy in China (Abbreviated Version)*, http://www.cresp.org.cn/uploadfiles/2/967/medium%20and%20long-term%20development%20plan%20for%20re%20in%20china%20eng.pdf

NEDO (New Energy and Industrial Technology Development Organisation) (2008) *Global Warming Countermeasures: Japanese Technologies for Energy Savings/GHG Emissions Reduction* (Revised Edition), NEDO, http://www.nedo.go.jp/content/100107259.pdf

NEDO (2009) *Profile of NEDO 2009–2010*, www.nedo.go.jp/kankobutsu/pamphlets/kouhou/profileofnedo2009_2010.pdf

New Energy Corporation (2010) *EnCurrent Vertical Axis Hydro Turbine New Energy Crop*, http://www.newenergycorp.ca/

NEV (Norges vassdrags-og energiedirektorat) (2008) *Vindkraftpotensialet utenfor norskekyten (offshore)* (K. Hofstad and L. Tallhaug), http://www.nve.no/global/publikasjoner/publikasjoner%202008/rapport%202008/rapport9-08.pdf

NFPA (Non Fossil Purchasing Agency) (2010) www.e-roc.co.uk/trackrecord.htm, accessed 17 February 2010

NORCOWE (Norwegian Centre for Offshore Wind Energy) (2010) http://www.norcowe.no/index.cfm

Nortrade.com (2009) *Wind Power: The Push Offshore*, 9 December, www.nortrade.com

NOWITECH (Norwegian Research Centre for Offshore Wind Technology) (2010) http://www.sintef.no/Projectweb/Nowitech/

NREL (National Renewable Energy Laboratory) (2004) *Future for Offshore Wind Energy in the United States* (W. Musial and S. Butterfield), http://www.nrel.gov/docs/fy04osti/36313.pdf

NREL (2010) *Large-Scale Offshore Wind Power in the United States: Assessment of Opportunities and Barriers*, www.osti.gov/bridge

NRG Bluewater Wind (2010) *Process & Timeline*, http://www.bluewaterwind.com/de_timeline.htm#agencies

N-RIP (National Renewables Infrastructure Plan) (2010) *National Renewables Infrastructure Plan*, Scottish Enterprise and Highlands and Islands Enterprise, http://www.scottish-enterprise.com/~/media/SE/Resources/Documents/Sectors/Energy/energy-renewables-reports/National-renewables-infrastructure-plan.ashx

Oceaneering (2010) *Medusa Spar Brochure*, http://www.oceaneering.com/oceandocuments/brochures/subseaproducts/MOPS%20-%20Medusa%20Spar.pdf

Ocean Power Delivery Ltd (2006) *World's First Wave Farm-Shipping of First Machine to Portugal*, Press release, 14 March 2006, http://www.oceanpd.com

OECD/IEA (Organisation for Economic Co-operation and Development and International Energy Agency) (2008) *Deploying Renewables: Principles for Effective Policies*, http://www.iea.org/w/bookshop/add.aspx?id=337, accessed 21 October 2010

Offshore Moorings (2009) *Ice Engineering: Moorings*, http://www.offshoremoorings.org/Moorings/2009/Group02_Prabhakar/OffshoreMooringsWEBSITE25sept2009/Moorings_Main_Page.htm

Offshore Moorings (2010) *Moorings*, http://www.offshoremoorings.org/moorings

Offshore-Technology.com (2010) *SPT Offshore*, http://www.offshore-technology.com/contractors/design-engineering-construction/spt_2/spt_24.html

Offshore Valuation Group (2010) *The Offshore Valuation*, p72, http://www.offshorevaluation.org

Offspring International (2010) *Polyester Rope Mooring*, http://www.offspringinternational.com/caseStudy/14/Polyester_Rope_Mooring_for_Thunder_Hawk_DeepDraft_Semi.htm

OJEU (Official Journal of the European Union) (2009) *Directive 2009/28/EC of the European Parliament and of the Council of 23 April 2009 on the Promotion of the Use of Energy from Renewable Sources and Amending and Subsequently Repealing Directives 2001/77/EC and 2003/30/EC*, Brussels

OPD (Ocean Power Technologies) (2006) *Making Waves in Power*, www.oceanpowertechnologies.com

Openhydro (2010) *Open Centre Turbine*, http://www.openhydro.com/development.html

OQPS (2010) *Marine (Scotland) Act 2010* (asp 5), http://www.oqps.gov.uk/legislation/acts/acts2010/pdf/asp_20100005_en.pdf

Oyster Aquamarine Power (2010) http://www.aquamarinepower.com/

Pelamis Wave Power (2010) *Installation*, http://www.pelamiswave.com/our-technology/installation

Pengfei, S. (2008) *Wind Energy Resource in China and Planning of Wind Farm Development*, Hydropower Planning General Institute, State Power, http://www.cwpc.cn/cwpc/en/technology/wind-resources, accessed 16 February 2010

Petersen, I. K., Kjaer Christensen, T., Kahlert, J., Desholm, M. and Fox, A. (2006) *Final Results of Bird Studies at the Offshore Wind Farms at Nysted and Horns Rev*, Denmark, National Environmental Research Institute, http://www.vattenfall.com/en/file/69662_Horns_Rev__Nysted_birds_7842547.pdf

PFI Market Intelligence (2009) *The Offshore Wind Revolution: Making Sense of the Opportunity*, Thomas Reuters, London

Polaski, K. (2003) 'Waiting for the waves', *IEA Ocean Energy Systems Newsletter*, issue 2, September 2003, http://www.iea-oceans.org/newsletters/news2.pdf

Previsic, M., Moreno, A., Bedard, R., Polagye, B., Collar, C., Lockard, D., Toman, W., Skemp, S., Thornton, S., Paasch, R., Rocheleau, R., Musial, W. and Hagerman, G. (2009) 'Hydrokinetic energy in the United States: Resources, challenges and opportunities', in *Proceedings of the 8th European Wave and Tidal Energy Conference*, Uppsala, Sweden, http://www.see.ed.ac.uk/~shs/Wave%20Energy/EWTEC%202009/EWTEC%202009%20(D)/papers/178.pdf

RAMBØLL (1999) *Kortlægning af bølgeenergiforhold i den danske del af Nordsøen*, http://www.waveenergy.dk/files/rep50105.pdf

RAVE (Research at Alpha Ventus) (2010) http://rave.iset.uni-kassel.de, accessed 24 August 2010

Reed, M. C. (2010) 'US Department of Energy Water Power Program', Presentation at SuperGen Marine Energy '10, 17 November 2010, www.supergen-marine.org.uk/drupal/content/past-evnts/assembly-2010

Rees, J. Larcombe, P., Vivian, C. and Judd, A. (2006) *Scroby Sands Offshore Wind Farm: Coastal Processes Monitoring*, Final Report for the Department of Trade and Industry, Cefas Lowestoft Laboratory, http://www.cefas.co.uk/media/326331/ae0262-final-report-scroby-owf.pdf

REN 21 (2009) *Recommendations for Improving the Effectiveness of Renewable Energy Policies in China*, http://www.ren21.net/pdf/Recommendations_for_RE_Policies_in_China.pdf

REN 21 (2010) *Global Status Report*, http://www.ren21.net/REN21Activities/Publications/GlobalStatusReport/tabid/5434/Default.aspx

RENBAR (2010) *Good Practices for Solving Environmental, Administrative and Socio-Economic Barriers in the Deployment of Renewable Energy Systems*, IEA–RETD report, http://www.iea-retd.org/files/ToR_RENBAR_090710.pdf

Renewable UK (2009) *UK Offshore Wind: Charting the Right Course*, Garrad Hassan study, http://www.bwea.com/pdf/publications/ChartingtheRightCourse.pdf

Renewable UK (2010) http://www.bwea.com/safety/index.html, and links within to the various specific reports and guidelines, accessed 21 September 2010

Risø (2010) *DTU's Wind Power Meteorology Programme and Energy & Environmental Data (EMD)*, Aalborg, http://www.emd.dk/Documentation/DK%20Wind%20Resource%20Map/

Risø National Laboratory for Sustainable Energy (undated) http://www.windatlas.dk/World/Index.htm

Rolls Royce (2010) *Platform Mooring Systems*, http://www.rolls-royce.com/marine/products/deck_machinery/pms/

SBM Atlantia (2010) *FPSO*, http://www.sbmatlantia.com/products/floating-solutions/fpsos/

Schuman, S. (2010) *China Renews Its Commitment to Renewable Energy*, http://switchboard.nrdc.org/blogs/bfinamore/china_renews_its_commitment_to.html

Scottish Enterprise (2010) *National Renewables Infrastructure Plan: Stage 2*, http://www.scottish-enterprise.com/your-sector/energy/energy-background/energy-reports/energy-renewable-energy-reports.aspx

Scottish Executive (2009) *National Renewables Infrastructure Plan*, Scotland

Scottish Government (2010a) *Marine (Scotland) Act*, http://www.scotland.gov.uk/Topics/marine/seamanagement/marineact, accessed 21 October 2010

Scottish Government (2010b) *Marine Scotland*, http://www.scotland.gov.uk/About/Directorates/Wealthier-and-Fairer/marine-scotland, accessed 21 October 2010

Scottish Government (2010c) *Saltire Prize Challenge*, http://www.scotland.gov.uk/Topics/Business-Industry/Energy/Action/leading/saltire-prize, accessed 11 February 2010

SEAI (Sustainable Energy Authority of Ireland) (2010) www.sei.ie/Renewables/Ocean_Energy, accessed 12 February 2010

SEEWEC (Sustainable Economically Efficient Wave Energy Converter) (2010) http://www.seewec.org/

SEI (2001) *Offshore Electricity Generating Stations: Note for Intending Developers*, http://www.sei.ie/uploadedfiles/RenewableEnergy/policydocument.pdf

SEI (2004) *Tidal and Current Energy Resources in Ireland*, www.seai.ie/Publications/Renewables_Publications/Tidal_Current_Energy_Resources_in_Ireland_Report.pdf

SEI (Sustainable Energy Ireland) (2005) *Ocean Energy in Ireland*, Department of Communications, Marine and Natural Resources, http://www.marine.ie/NR/rdonlyres/86491414-3E7E-48E5-A0E1-287CA9191C61/0/OceanEnergyStrategy.pdf

Siemens AG (2009) *Oceans of Opportunities Renewable Energy Division*, Order no E50001-W310-A118-X-4A00

Siemens AG (2010) *Siemens Wind Turbine SWT-3.6–107*, http://www.energy.siemens.com/hq/en/power-generation/renewables/wind-power/wind-turbines/swt-3-6-107.htm, accessed 10 February 2010

Simoes, V. C., Rammer, C., Sellenthin, M. O. and Thorwarth, S. (2007) *Monitoring and Analysis of Policies and Public Financing Instruments Conducive to Higher Level of R&D Investments: The 'Policy Mix' Project, Country Review Portugal*, http://ec.europa.eu/invest-in-research/pdf/download_en/portugal.pdf

SMRU Limited (2010) *Approaches to Marine Mammal Monitoring at Marine Renewable Energy Developments: Final Report*, www.thecrownestate.co.uk/marine_mammal_monitoring.pdf

South West RDA (2010) *Wave Hub Successfully Deployed on Ocean Floor*, http://www.southwestrda.org.uk/news_and_events/2010/september/wave_hub_successfully_deployed.aspx, accessed 8 September 2010

Spanish Ministry of Environment and Rural and Marine Areas (2009) Report, 20 April 2009, http://www.mapa.es/gabinete/nota.asp?codi=22222_AH200409, accessed 16 March 2010

Stingray Assembly Engineering Business Ltd (2005) *Oscillating Hydrofoil*, http://www.esru.strath.ac.uk/EandE/Web_sites/05-06/marine_renewables/technology/oschydro.htm

Supergen Marine (2010) Supergen Marine Energy Research Consortium, http://www.supergen-marine.org.uk/drupal/content/publications

Sway (2010) http://www.sway.no/publish_files/Pressemelding_Enova.pdf

Swedish Energy Agency (2009a) *Swedish Energy Research 2009*, http://www.energimyndigheten.se/en/Research/

Swedish Energy Agency (2009b) *The Electricity System, 2009*, http://www.energimyndigheten.se/en/Press/News/New-publication-The-electricity-certificate-system-2009/

Takagi, T., Fukuda, H., Tanikawa, R. and Quuki, Y. (2006) 'Asian wind power resource map', in *Proceedings of European Wind Energy Conference and Exhibition*, Athens, CRC Solutions Corp. Inc., Japan

The Crown Estate (2009a) *Guide to an Offshore Wind Farm*, http://www.thecrownestate.co.uk/guide_to_offshore_windfarm.pdf

The Crown Estate (2009b) *Towards Round 3: Building the Offshore Wind Supply Chain*, http://www.thecrownestate.co.uk/round3_supply_chain_gap_analysis.pdf

The Crown Estate (2010a) *World's First Wave and Tidal Energy Leasing Round*, http://www.thecrownestate.co.uk/newscontent/92-pentland-firth-developers.htm

The Crown Estate (2010b) *Round 3 Announcements*, http://www.thecrownestate.co.uk/round3-announcements, accessed 11 February 2010

The World of Wind Atlases (2010) *Wind Atlases of the World*, http://www.windatlas.dk/World/Index.htm, accessed 12 February 2010

Thorpe, T. (1992) *A Review of Wave Energy*, vols 1 and 2, ETSU Report no R-72, http://uk.linkedin.com/in/thomaswthorpe

Thorpe, T. (1999) *A Brief Review of Wave Energy*, ETSU, http://www.mech.ed.ac.uk/research/wavepower/Tom%20Thorpe/Tom%20Thorpe%20report.pdf

UKERC (UK Energy Research Centre) (2008) *Marine (Wave and Tidal Current) Renewable Energy Technology Roadmap*, UKERC, UK

UKERC (2009) *Decarbonising the UK Energy System: Accelerated Development of Low Carbon Energy Supply Technologies*, UKERC Energy 2050 Research Report no 2, UK

UKERC and University of Edinburgh (2008) *UKERC Marine (Wave and Tidal Current) Renewable Energy Technology Roadmap*, UKERC, UK

University College Cork (UCC) (2009) *Non-Technical Barriers to Wave Energy Development: Comparing Progress in Ireland and Europe* (G. Dalton, N. Rousseau et al), University College Cork, Cork, Ireland

Upwind (2010) www.upwind.eu, accessed 11 February 2010

US Department of the Interior (2006) *Wave Energy Potential on the U.S. Outer Continental Shelf*, Renewable Energy and Alternate Use Program, May 2006, US

Van Hulle, F., Cabooter, Y., Lebot, S., Van Lancker, V., Deleu, S., Driessen, J. and Soens, J. (2004) *Prospects for Offshore Wind Energy from the Belgian Continental Shelf*, http://2004ewec.info/files/24_1400_fransvanhulle_01.pdf

Vestas (2010) http://www.vestas.com/en/wind-power-solutions.aspx, accessed 15 February 2010

Voith Hydro (undated) http://pesn.com/2007/08/07/9500489_VAOT_v_HAOT/Wando_HAOT.jpg

WAB (2009) *Offshore Wind Energy Magazine*, issue 2, p56ff, September 2009, Bremerhaven, Germany

Watson, Farley and Williams (2008) *Green Certificates Regime as Amended by Budget Law 2008*, http://www.wfw.com/Publications/Publication459/$FILE/Italy's%20Green%20Certificates%20Regime%202008.pdf

Wave Energy Centre (2004) *Potential and Strategy for the Development of Wave Energy in Portugal*, WaveEC, http://www.wavec.org/client/files/Summary_DGGE_ingl.pdf

Wave Energy Centre (2010a) *WaveEC*, http://www.wavec.org

Wave Energy Centre (2010b) *OWC Pico Power Plant*, http://www.pico-owc.net

Wavenet (2003) *Results from the Work of the European Thematic Network on Wave Energy*, ERK-CT-1999-2001, European Community, March (2003), http://www.wave-energy.net/Library/WaveNet%20Full%20Report(11.1).pdf.

Waveplan (2009) *State of the Art Analysis*, http://www.waveplam.eu/files/downloads/SoA.pdf

Webb, R. (2009) 'First osmosis power plant goes on stream in Norway', *New Scientist*, 26 November, http://www.newscientist.com/article/dn18204-first-osmosis-power-plant-goes-on-stream-in-norway.html, accessed 16 February 2010

WEC (World Energy Council) (2007) *Survey of Energy Resources*, http://www.worldenergy.org/documents/ser2007_final_online_version_1.pdf

WEC (2009) *Survey of Energy Resources Interim Update 2009*, http://www.worldenergy.org/publications/survey_of_energy_resources_interim_update_2009/default.asp

Wikipedia (2010) en.wikipedia.org/wiki/Floating_wind_turbine

Windpower Monthly (2008) 'Portugal's offshore wind potential', *Windpower Monthly*, 1 November

Windpower Monthly (2010) *Husum Special Report: Technology Latest*, Supplement to *Windpower Monthly*, September, p21

WRI (World Resources Institute) (2009) *It Should be a Breeze Harnessing the Potential of Open Trade and Investment Flows in the Wind Energy Sector*, Working Paper Series, WP 09–14 (J. Kiregaard, T. Hanemann et al), WRI

Zentech International Ltd (2010) http://www.zentech.co.uk/images/paper08_freehang.gi

Appendix B
List of Acronyms and Abbreviations

ABS American Bureau of Shipping
AC alternating current
ADEME French Environment and Energy Management Agency
ADORET Acceleration the Deployment of Offshore Renewable Energy
AIS automatic identification systems
BOE MOEA Taiwan Bureau of Energy, Ministry of Economy Affairs
BOP balance of plant
BSH Bundesamt für Seeschifffahrt und Hydrographie
BTU British thermal unit
CALM catenary anchor leg
CAPEX capital expenditure
CCGT combined cycle gas turbine
CDM Clean Development Mechanism
CDTI Spanish Centre for the Development of Industrial Technology
CEER Centre for Environment-Friendly Energy Research
CMA China Meteorological Association
CNOOC China National Offshore Oil Corp
CO_2 carbon dioxide
COE cost of energy
COWRIE Collaborative Offshore Wind Research into the Environment
CP compliance period
DC direct current
DECC UK Department of Energy and Climate Change
DEMP Danish Environmental Monitoring Programme
DNREC Delaware Department of Natural Resources and Environmental Control
DNV Det Norske Veritas
DOE Department of the Environment
DSCR debt service coverage ratio
DTI UK Department of Trade and Industry
EC European Commission

ECA European Credit Agency
EDA Electricidade dos Acores
EDP Electricidade de Portugal
EEG German Renewable Energy Sources Act
EEZ Exclusive Economic Zone
EIA environmental impact assessment
EIS environmental impact statement
EKF Eksport Kredit Fonden
EMEC European Marine Energy Centre
EMF electromagnetic field
EPA US Environmental Protection Agency
EPC engineering procurement and construction
EPRI US Electric Power Research Institute
ES environmental statement
ETI Energy Technology Institute
EU European Union
EU ETS European Union Emissions Trading System
EU Skills Energy & Utility Skills
EWEA European Wind Energy Association
FERC Federal Energy Regulatory Commission
FOAK first-of-a-kind
FORCE Fundy Ocean Research Centre for Energy
FP7 Framework Programme 7
FIT feed-in tariff
GDP gross domestic product
GHG greenhouse gas
GL Germanischer Lloyds
GTRC Georgia Tech Research Corporation
GW gigawatt
GWh gigawatt hours
H&S health and safety
HAZID hazard identification
HSE Health and Safety Executive
HSMP Health and Safety Management Plan
HVAC high voltage alternative current
HVDC high voltage direct current
IEA–RETD International Energy Agency–Renewable Energy Technology Development
IEC International Electrotechnical Commission
IFI International Finance Institution
IMO International Maritime Organization
INETI National Institute of Engineering and Industrial Technologies

IPC Infrastructure Planning Commission
IRR internal rate of return
IST Instituto Superior Técnico
ITC investment tax credit
ITRI Industrial Technology Research Institute
JAMSTEC Japan Agency for Marine-Earth Science and Technology
kW kilowatt
kWh kilowatt hour
kV kilovolt
LCR local content requirement/rule
MCT Marine Current Turbine
MHK marine and hydrokinetic
MI major intervention
MM Mott MacDonald
MMO Marine Management Organization
MMS mandatory market share
MMS Minerals Management Service
MPA Marine Protected Area
MRDF Marine Renewable Deployment Fund
MRPF Marine Renewables Proving Fund
MW megawatt
MUMM Belgian Management Unit of the North Sea Mathematical Models
NASA US National Aeronautics and Space Administration
NDRC National Development and Reform Commission
NEDO New Energy and Industrial Technology Development Organization
NGO non-governmental organization
NIRO Renewables Obligation (Northern Ireland)
nm nautical mile
NO_x nitrogen oxide
NOAA National Oceanic and Atmospheric Administration
NOAK nth-of-a-kind
NORCOWE Norwegian Centre for Offshore Wind Energy
NOWITECH Norwegian Research Centre for Offshore Wind Technology
NPV net present value
NREL US National Renewable Energy Laboratory
NTS non-technical summary
NVE Norges vassdrags-og energidirektorat
O&M operation and maintenance
OCS Outer Continental Shelf
OE ocean energy
OFTO offshore transmission
OPEX operational expenditure

OPT Ocean Power Technologies
OTEC ocean thermal energy converter/conversion
OWC oscillating water column
PLC programme logic control
PPA power purchase agreement
PPE personal protective equipment
PPP public–private partnership
PSHSS pitching/surging/heaving/sway/sloped
PTC production tax credit
R&D research and development
PV photovoltaic
RAVE research at Alpha Ventus
RD&D research, development and demonstration/deployment
REFIT renewable energy feed-in tariff
REN Redes Energéticas Nacionais
RENBAR IEA–RETD 'Good practices for solving environmental, administrative and socio-economic barriers in the deployment of renewable energy systems'
RETD renewable energy technology development
RFI request for information
RFP request for proposal
RO Renewables Obligation
ROC renewable obligation certificate
ROS Renewables Obligation (Scotland)
ROV remote operating vehicle
rpm revolutions per minute
RPS renewable portfolio standard
RTD Research Trade and Development programme
SA Sociedade Anónima
SALM single anchor leg
SDE Subsidies Duurzame Energie
SEA strategic environmental assessment
SEAI Sustainable Energy Authority of Ireland
SEEWEC Sustainable Economically Efficient Wave Energy Converter
SER Syndicat des Energies Renouvelables
SGPS Sociedade Gestora de Participações Sociais
SME small- and medium-sized enterprise
SNH Scottish Natural Heritage
SO_x sulphur oxide
SO_2 sulphur dioxide
SPAR single point + reservoir
SPV special purpose vehicle
SSC Sector Skills Council

ToR terms of reference
TPWind Wind Energy Technology Platform
TRL technology readiness level
TSO transmission system operator
TW terawatt
TWh terawatt hours
UKCES UK Commission for Employment and Skills
UNCLOS United Nations Convention on the Law of the Sea
USACE United States Army Corps of Engineers
UTC United Technologies
VAT value-added tax
WaveEC Wave Energy Centre
WEC wave energy converter
WTG wind turbine generator

Appendix C
Assumptions Underlying the Sensitivity Analysis

		Offshore Wind	Wave/ Tidal
Technical assumptions			
Generation capacity	MW	200	50
Average annual energy generation	GWh	700	131.4
Capacity factor	%	40	30
Manufacture			
WTG	€ million	320.0	167.5
Foundations	€ million	62.5	77.5
Cables	€ million	82.5	87.5
Substations	€ million	100.0	50.0
Total manufacture	€ million	565.0	382.5
Vessels			
WTG/foundations	€ million	37.5	37.5
Cables	€ million	26.3	26.3
Other	€ million	0.0	0.0
Total vessels	€ million	63.8	63.8
Project management	€ million	15.0	20.0

		Offshore Wind	Wave/ Tidal
Design and supervision			
Development costs	€ million	20.0	20.0
Insurance	€ million	15.0	20.0
Total design and supervision	€ million	35.0	40.0
Contingency	€ million	94.3	89.3
Total capital cost	€ million	773.1	595.5
O&M costs (relative to CAPEX)			
OPEX	% of construction CAPEX/year	3.0%	5.0%
Insurance	% of construction CAPEX/year	0.5%	0.5%
O&M costs (absolute)			
OPEX	€ million/year	18.9	22.3
Insurance	€ million/year	3.2	2.3

Appendix D
Exchange Rates

Currency	UK £	Euros (€)
UK€1 =	1.00	1.18
€1 =	0.85	1.00
US$1 =	0.64	0.75
Cdn$1 =	0.63	0.75
1 Danish krone =	0.11	0.13
1 Japanese yen =	0.0077	0.0091

Source: www.xe.com, last accessed 6 December 2010

Appendix E
Country Profiles

E.1 Introduction

This appendix outlines the country profiles of the 18 countries in terms of the current industry context. The list of countries is shown in Table E.1.

Table E.1 Countries covered by the study

RETD Countries	Other Countries of Interest
Canada	Belgium
Denmark	Finland
France	Spain
Germany	Sweden
Ireland	US
Italy	China
Japan	Taiwan
The Netherlands	Portugal
Norway	
UK	

For each country, an overview of the following areas is provided, including:

- offshore wind resource;
- wave resource;
- tidal resource;
- processes for permitting, licensing and allocating seabed rights;
- support mechanisms for project deployment;
- support mechanisms for development of technologies;
- other support initiatives (where applicable); and
- deployment.

The references given correspond with the full reference list provided in the book.

E.2 Canada

E.2.1 Offshore wind resource

Canada has a good coastal and offshore wind resource, as shown in Figure E.1. The near-shore wind resource is characterized by a mean wind speed of 6m/s and above. Offshore wind speed is approximately 8m/s and above. Wind speeds at the southern regions of the Atlantic and Pacific coasts tend to be lower than in the north.

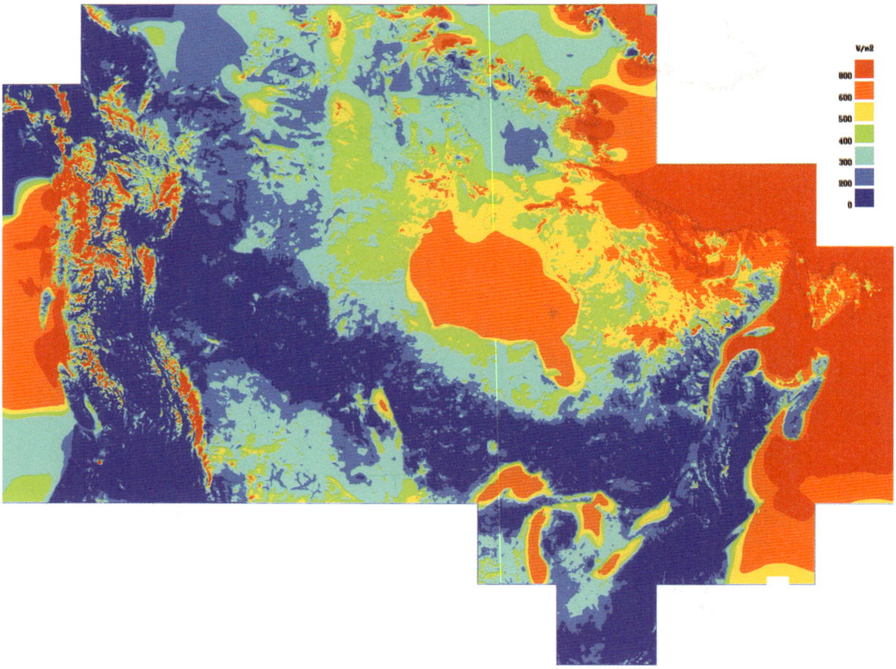

Figure E.1 Mean wind speed 50m above ground (*Canadian Wind Energy Atlas*)
Source: © Source: Environment Canada, Canadian Wind Energy Atlas, 2003

E.2.2 Wave resource

Assessments of the yearly wave energy flux suggest that Canada has a relatively intense wave climate at both the west and east coast. Depending upon the locations and the distance from shore, 40 to 60kW/m can be achieved.[1] Monthly variations of the wave regime have been analysed by the Canadian Hydraulics Centre, drawing on a range of measurement positions in water depth greater than 200m. These assessments suggest that wave energy is highest during winter times (November to March) and lowest during the summer months. The distribution of wave energy in water depth shallower than 200m is noted to be poorly assessed to date.[2]

Figure E.2 shows the distribution of wave energy at the east and west coast of Canada. The mean annual wave energy flux at exposed sites in deep water off the eastern coast is in the range of 45 to 55kW/m. Near shore, the mean power available is in the order of 30 to 45kW/m. At the east coast, the energy flux is in the range of 42 to 45kW/m, with a near-shore intensity of around 25 to 30kW/m. The annual mean wave power off Canada's Pacific Coast, along the 1000m isobaths, is estimated to be around 37GW. Equally, for Canada's Atlantic Coast, around 146GW is estimated.[3]

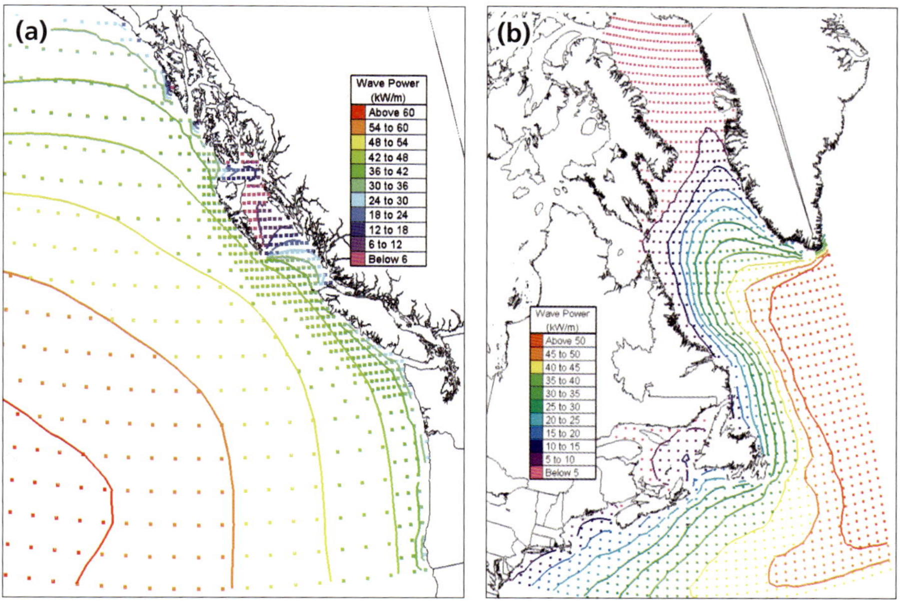

Figure E.2 Mean annual wave power for Canada's coasts: (a) Mean annual wave power in the north-east Pacific derived from WW3-ENP hindcast data; (b) mean annual wave power derived from AES40 hindcast data

Source: Cornett, A.M. (2006) Inventory of Canada's Marine Renewable Energy Resources. Report CHC-TR-041, Canadian Hydraulics Centre, National Research Council, Ottawa

E.2.3 Tidal resource

While wave energy is available throughout the ocean in varying intensity, tidal energy varies substantially and its force is amplified by bathymetry and coastal features. The maps of the east and west coast of Canada shown in Figure E.3 indicate some sites with good to high energy potential for a tidal energy scheme. According to assessments undertaken, Canada has a potential tidal current energy resource of 42,240MW.[3]

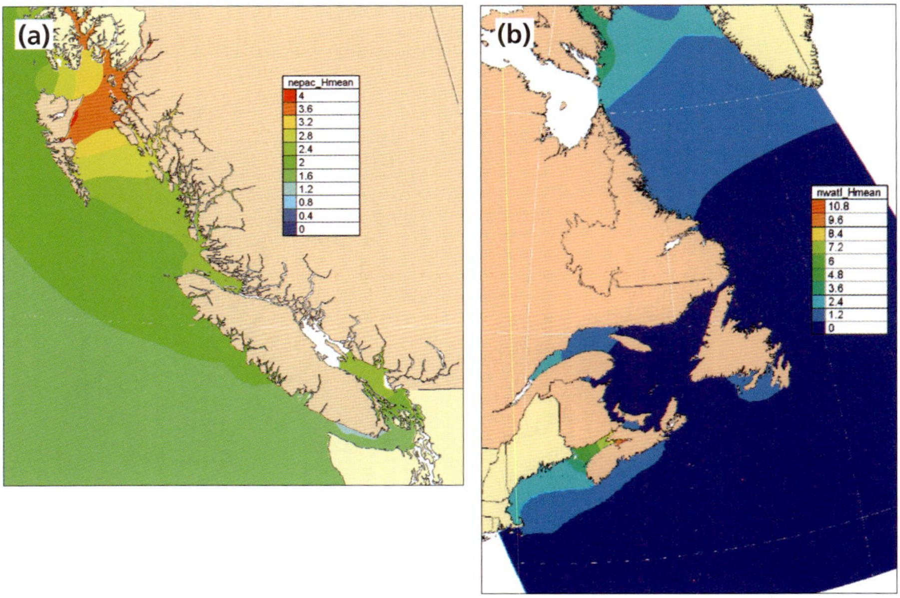

Figure E.3 Mean tidal range along Canada's coast: (a) Mean tide range along the British Columbia coast (m); (b) mean tide range in eastern Canada
Source: Cornett, A.M. (2006) Inventory of Canada's Marine Renewable Energy Resources. Report CHC-TR-041, Canadian Hydraulics Centre, National Research Council, Ottawa

A number of potential sites with a size of above 1MW are shown in Figure E.4. Overall, the realisation of all 191 proposed schemes with an average size of 221MW would lead to a total capacity of 42,240MW.[4]

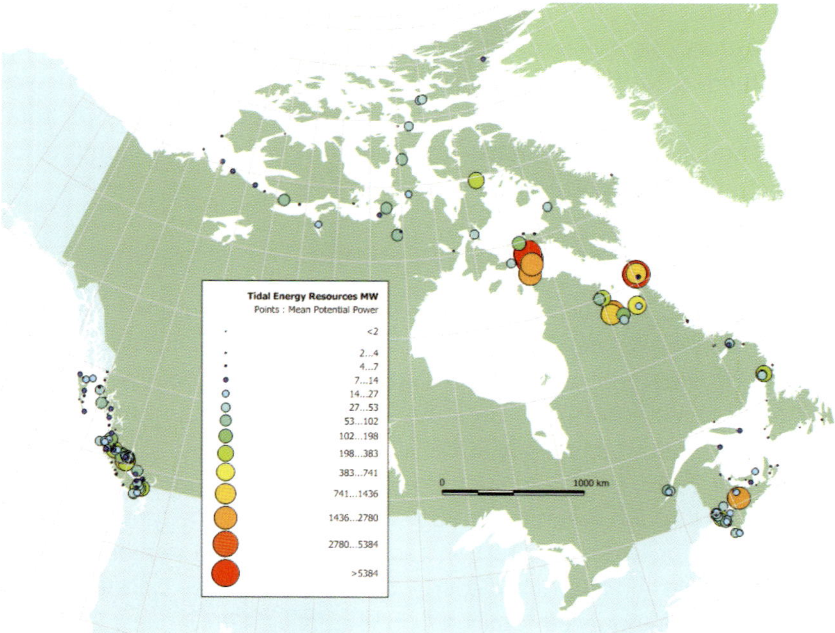

Figure E.4 Leading tidal current power sites, Canada

Source: Cornett, A.M. (2006) Inventory of Canada's Marine Renewable Energy Resources. Report CHC-TR-041, Canadian Hydraulics Centre, National Research Council, Ottawa

E.2.4 Government bodies for permitting, licensing and allocating seabed rights

Fisheries and Oceans Canada collaborates with other federal departments and coastal provinces regarding ocean energy developments.[5]

For offshore renewable energy projects located within provincial boundaries, the provinces have primary responsibility for permitting and managing projects (e.g. land tenure, electricity regulation). The federal government is responsible for regulating activities under its jurisdiction (e.g. fisheries, navigable waters).

Offshore renewable energy projects located outside the boundaries of provinces are subject to a federal land tenure authorization, which is administered by Natural Resources Canada.

E.2.5 Support mechanisms for project deployment

Canada has an incentive payment scheme at federal level. The ecoENERGY for Renewable Power programme was established in 2007 and provides an incentive of Cdn$1 cent/kWh (€0.64 cent/kWh) for up to ten years for electricity generated from eligible renewable energy projects commissioned between 1 April 2007 and 31 March 2011. Some individual provinces have also legislated renewable portfolio standards, use procurement mechanisms, such as calls for power and requests for proposals, while others have established or intend to establish renewable feed-in tariff (FIT) programmes. Ontario's FIT programme was introduced in October 2009. The tariffs provided are intended to cover total project costs and provide a reasonable rate of return over a 20-year contract for most technologies, and 40 years for waterpower.[6]

E.2.6 Support mechanisms for development of technologies

The national ecoENERGY Technology Initiative is a Cdn$230 million investment in clean energy research, development and deployment (RD&D) programme to support the development of next-generation energy technologies.[7] It is a five-year programme which runs until 2012. It includes funding for offshore wind and ocean energy technology RD&D.

Introduced in 2009, the Clean Energy Fund aims to reduce greenhouse gas (GHG) emissions through green technology demonstration. The fund will invest Cdn$795 million for large-scale carbon capture and storage demonstrations, including renewable and alternative energy technologies. Three carbon capture projects have been announced totalling Cdn$466 million. In January 2010, 19 successful projects were announced in response to a call for proposals under the renewable and clean energy portion of the fund and up to Cdn$146 million will be invested in these projects over five years.[8] The first projects to be supported include the Bay of Fundy project in Nova Scotia, outlined in section E.2.8.

E.2.7 Other support initiatives

Another initiative that supports the offshore energy sector in Canada is the *Ocean Energy Atlas* project, which was established in 2005 after recognition that development of a comprehensive assessment of Canada's marine resources was a fundamental step towards the development of ocean energy in Canada. Funding for the first year was provided by the Technology and Innovation Research and Development Programme, administered by Natural Resources Canada. The project maps, at a regional scale on an interactive digital website, Canada's renewable ocean energy resources.[9]

E.2.8 Deployment

North America's largest offshore wind project is proposed near British Columbia, totalling 1750GW and built over five phases. The first phase will comprise 110 3.6MW Siemens turbines.[10]

The Fundy Ocean Research Centre for Energy (FORCE), a public–private partnership in Nova Scotia, is developing a multi-million dollar tidal pilot demonstration project in the Minas Passage of the Bay of Fundy. This will be the first Canadian deployment of commercial-scale tidal turbines. It has been given

Cdn$70 million of funding and will comprise three large-scale (1MW) machines. The project was expected to reach completion in 2010.[11]

E.3 Denmark

E.3.1 Offshore wind resource

Figure E.5 shows the Danish wind resource map. It is apparent that Denmark has a considerable wind resource, with an average shoreline and offshore wind speed of above 7m/s. The Danish offshore wind resource is estimated to amount to 8000MW, or around 26TWh/year, with an additional 4000MW at water depth greater than 20m.[12]

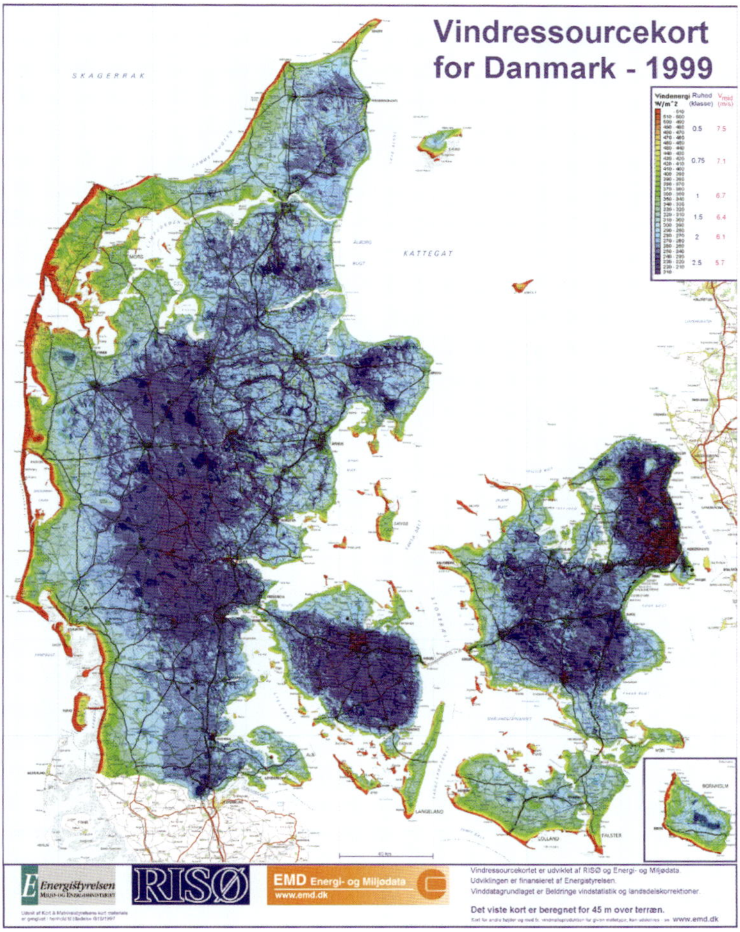

Figure E.5 Danish Wind Resource Map, 1999

Source: Risø DTU's Wind Power Meteorology Programme and Energy & Environmental Data (EMD), Aalborg, http://www.emd.dk/Documentation/DK%20Wind%20 Resource%20Map/

E.3.2 Wave resource

While Denmark is located in a relatively sheltered area of the North Sea, some locations in the north-western regions experience a relatively favourable wave regime with estimates of about 30TWh/year.[13] An assessment of specific sites in the Danish North Sea undertaken by RAMBØLL[14] shows that wave energy flux is strongly dependent upon water depth and therefore distance from shore. The wave energy flux at the selected sites varies between 7kW/m for sites at a water depth of 20m and 24kW/m at a depth of 71m. Measurement points are shown in Figure E.6.

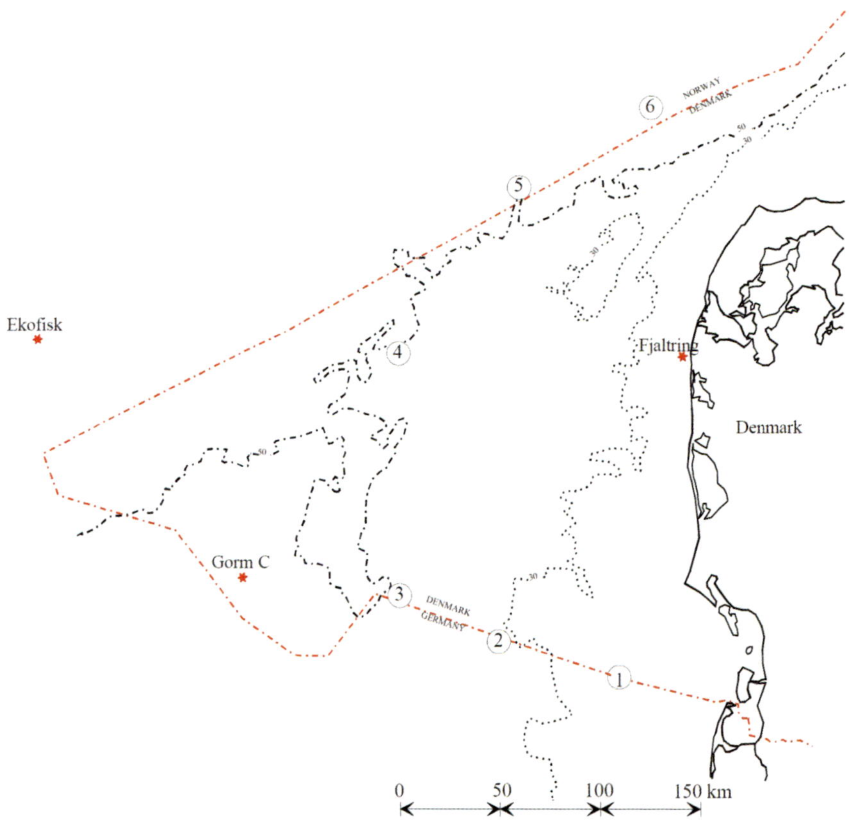

Figure E.6 Overview of assessed locations in the Danish North Sea
Source: RAMBØLL for the DEA (Danish Energy Agency), 1999

E.3.3 Tidal resource

No studies related to the Danish tidal resource have been available.

E.3.4 Processes for permitting, licensing and allocating seabed rights

The Danish Energy Agency is the competent authority for offshore wind power projects in Denmark. It is responsible for approving new projects, as well as carrying out research on environmental impacts and future locations. It cites itself as a 'one-stop shop' for offshore wind permits since it coordinates with other relevant authorities on the different permitting aspects. The permitting conditions for offshore farms are set by the Danish Electricity Supply Act of November 2008. There are two main ways to apply for an offshore wind permit: responding to a government call for tenders for development at specific sites, or through the 'open-door procedure' for other non-tendered locations.[15]

E.3.5 Support mechanisms for project deployment

Denmark's 2009 Promotion of Renewable Energy Act establishes FITs for renewable energy. Wind generation projects connected under the 'open-door procedure' receive a subsidy that amounts to around 0.273 Danish krone/kWh (€3.7 cents/kWh) on top of the electricity market price for power exported during the first 22,000 full load hours, which is equivalent to the order of eight years' operation at typical capacity factors and availabilities.[15]

Where offshore wind farms are developed following a call for tender, the state specifies the fixed electricity price in the call.[15] The wind farm at Horns Rev 2, for example, will be provided a FIT that, when added to the market price of electricity, will amount to 0.518 Danish krone/ kWh (~ €7 cents/kWh). For the wind farm at Rødsand 2, the total tariff (premium plus market price) will equal 0.629 Danish krone/kWh (~ €8.4 cents/kWh). A price supplement is also granted to cover payment of any feeding fee related to supplying electricity to the grid. The premiums apply to up to 10TWh of electricity generation (depending upon the specific contract terms) and for a maximum of 20 years after the wind farm has been connected to the grid.[16] In February 2008, the Danish parliament set a target for a further 400MW of offshore wind turbines to come online in 2012.[17]

For wave energy and other renewable technologies considered of strategic importance by the Danish government, the price received is 0.60 Danish krone/ kWh (€8 cents/kWh) for ten years from the date of grid connection, and 0.40 Danish krone/kWh (€5 cents/kWh) for the next ten years.[17]

E.3.6 Support mechanisms for development of technologies

In February 2008, the Danish government signed a broad-ranging energy agreement that ensures the development of renewable energy, increased energy efficiency and more research into energy technologies. However, limited publicly available information was found.

E.3.7 Deployment

In 2009, Denmark installed 98 offshore turbines, totalling 230MW across two separate wind farms. This brought the installed total by the end of 2009 to 305 turbines, totalling 639.15MW across nine wind farms.[18] Danish offshore wind farms are shown in Table E.2 and marine energy installations in Table E.3.

Table E.2 Danish offshore wind farms

Project	Number of Turbines/Capacity	Capacity
1 Vindeby (1991)	11 turbines	5MW
2 Tunø Knob (1995)	10 turbines	5MW
3 Middelgrunden (2000)	20 turbines	40MW
4 Horns Rev I (2002)	80 turbines	160MW
5 Rønland (2003)	8 turbines	17MW
6 Nysted (2003)	72 turbines	165MW
7 Samsø (2003)	10 turbines	23MW
8 Frederikshavn (2003)	3 turbines	7MW
9 Horns Rev II (2009)	91 turbines	209MW
10 Avedøre Holme (2009/10)	3 turbines	10–13MW
11 Sprogø (2009)	7 turbines	21MW
12 Rødsand II (2010)	90 turbines	207MW
13 Anholt (2012)	400MW	400MW

Source: *Danish Energy Agency (2010)*

Table E.3 Danish marine energy installations

Installation	Stage	Capacity (kW)	Date Commissioned
Wave Star	Operational	5.5	2006
Wave Dragon	Being tested	20	n.c.
Poseidon	Being tested	140	2008

Note: n.c = not commissioned.
Source: *EurObserv'ER (2010)*

Waveplane has also been tested at two separate sites in Denmark until August 2009. A DEXA Wave Energy model was also tested until October 2009 and an application for further testing until August 2011 is in process.[19]

E.4 France

E.4.1 Offshore wind resource

Figure E.7 shows that the Atlantic shores of France are exposed to a good wind climate, with wind speeds of around 8m/s at 50m heights. At the Mediterranean coasts, the local phenomenon of the Mistral winds is evident, which leads to a high wind resource particularly in comparison to other areas of the Mediterranean. Wind speeds at 50m heights can exceed 8m/s. France's offshore wind resource at a water depth below 40m and a distance from shore below 30km has been estimated to amount to 13,000MW and 44TWh/year, respectively.[20]

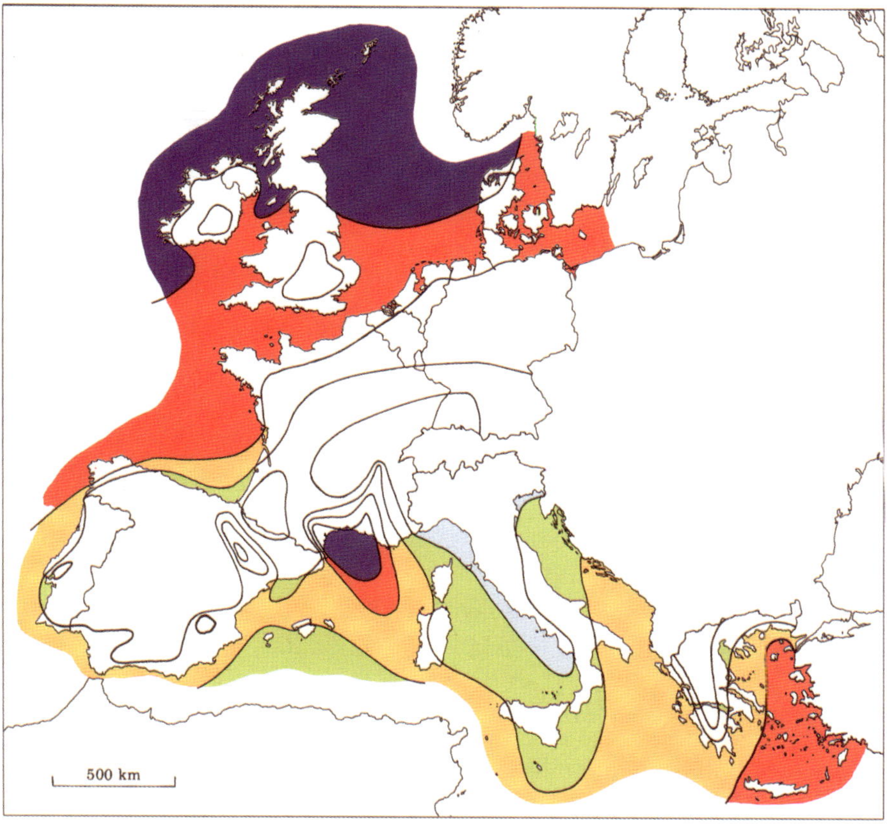

Wind resources over open sea (more than 10 km offshore) for five standard heights										
	10 m		25 m		50 m		100 m		200 m	
	$m s^{-1}$	$W m^{-2}$	$m s^{-1}$	$W m^{-2}$	$m s^{-1}$	$W m^{-2}$	$m s^{-1}$	$W m^{-2}$	$m s^{-1}$	$W m^{-2}$
	> 8.0	> 600	> 8.5	> 700	> 9.0	> 800	> 10.0	> 1100	> 11.0	> 1500
	7.0-8.0	350-600	7.5-8.5	450-700	8.0-9.0	600-800	8.5-10.0	650-1100	9.5-11.0	900-1500
	6.0-7.0	250-300	6.5-7.5	300-450	7.0-8.0	400-600	7.5- 8.5	450- 650	8.0- 9.5	600- 900
	4.5-6.0	100-250	5.0-6.5	150-300	5.5-7.0	200-400	6.0- 7.5	250- 450	6.5- 8.0	300- 600
	< 4.5	< 100	< 5.0	< 150	< 5.5	< 200	< 6.0	< 250	< 6.5	< 300

Figure E.7 European wind resource over open sea

Source: World of Wind Atlases, European section www.windatlas.dk copyright Risø DTU

E.4.2 Wave resource

Annual power levels at the Atlantic Coast are suggested to range between 30 to 40kW/m and up to 5kW/m in the Mediterranean Sea.[21] The total resource is estimated to be approximately 28GW.[22]

E.4.3 Tidal resource

While favourable locations might be present at the Atlantic Coast of France, no specific research has been identified.

E.4.4 Processes for permitting, licensing and allocating seabed rights

The framework for undertaking offshore economic activity is not adapted to wind energy, and there is no specific legislative or administrative framework for offshore wind energy development in France. This has led to long delays to the first proposed offshore wind farm announced in 2005 (see below). Work has begun, however, on simplifying offshore planning procedures in this regard.[23]

E.4.5 Support mechanisms for project deployment

Since 2001, France has promoted renewable energy projects through a feed-in tariff mechanism and a tender system for large projects. For offshore installations, there is a fixed tariff of €13 cents/kWh for the first ten-year period. For the following ten years, the tariff is set at €3 cents/kWh for 3900 hours (and more) and at €13 cents/kWh for 2800 hours (or less), with a sliding scale for amounts in between.[24] For ocean energy, the tariff is fixed at €15 cents/kWh for 20 years. Under the tender system, one offshore wind project (105MW) was announced in 2005.[25]

In 2007, the French Syndicat des Energies Renouvelables (SER) suggested an offshore wind power capacity target of 6GW by 2020.[26]

E.4.6 Support mechanisms for development of technologies

In July 2008, a demonstration fund was created by three government ministries to provide €400 million over four years to help finance the development of energy technologies for which there currently is no viable market. The fund is a joint effort between the Ministry of Ecology, Energy, Sustainable Development and Territorial Planning, the Ministry of Higher Education and Research and the Ministry of Economy, Industry and Employment, and is managed by the French Environment and Energy Management Agency (ADEME). The fund targets manufacturers or public–private partnerships in need of financing for high-cost demonstration projects with experimental goals, for which market prospects are long term.[27]

In July 2009, the fund issued a call for proposals in offshore energy demonstration projects. The call was open to tidal and wave, offshore wind and ocean thermal projects. It closed on 15 October 2009.[28] As of February 2010, the results had not been released.

E.4.7 Deployment

Due to permitting problems delaying its first planned offshore wind farm, France currently has no offshore wind installations. The only non-barrage project includes a 10kW Hydro-Gen marine device being tested since 2008.[29]

EDF, the leading electricity producer in France, plans a pilot tidal turbine system off the coast of Brittany. The project, consisting of four to ten turbines, with a total capacity of between 2MW and 4MW, will be sited at Paimpol-Bréhat (Côtes d'Armor). In October 2008, EDF stated that the company had appointed OpenHydro of Ireland to equip the demonstration tidal farm, which is scheduled to be connected to the grid from 2011 onwards.[30]

E.5 Germany

E.5.1 Offshore wind resource

Germany's coasts both in the North and Baltic Sea are exposed to considerable wind speeds above 8m/s at 50m heights (see Figure E.7). Estimates of the theoretical German offshore wind resource depend upon the assumption of feasible water depth and distance from shore. The following estimates have been made[31]:

- up to 10m water depth and 10km distance from shore = 20TWh/year;
- up to 20m water depth and 20km distance from shore = 130TWh/year; and
- up to 30m water depth and 30km distance from shore = 200TWh/year.

Other estimates considering locations up to 40m water depth and below 30km distance from shore amount to an offshore resource of 13,000MW and 45TWh/year.[32]

E.5.2 Wave resource

Due to the sheltered location in the North Sea, the German coast has a poor wave resource with below 10kW/m wave energy flux.

E.5.3 Tidal resource

The North Sea coast of Germany experiences tidal variations that have led to the formation of the Wadden Sea, which is a strongly protected ecosystem. No tidal resource assessments regarding the potential energy output can be referenced.

E.5.4 Processes for permitting, licensing and allocating seabed rights

The Federal Emission Control Act governs the licensing and construction of wind farms in coastal waters (up to 12 nautical miles from the coastline). The licensing procedure must be carried out by the competent emission control authorities in the relevant coastal states, including an environmental impact assessment and public participation. The construction and operation of offshore wind farms in the Exclusive Economic Zone (EEZ), which comprises the marine territory beyond the coastal waters, is based on the Federal Maritime Responsibilities Act in conjunction with the Offshore Installations Ordinance. These state that the construction, operation and significant alteration of fixed or floating structural or technical equipment for the generation of energy from wind must be licensed by the Federal Maritime and Hydrographic Agency (BSH).[33]

E.5.5 Support mechanisms for project deployment

Under the new Renewable Energy Sources Act (EEG), implemented in 2009 and superseding the 2004 Act, electricity from offshore wind farms operational from January 2009 onwards has been guaranteed a feed-in tariff for 20 years at the levels shown in Table E.4. The rate will be kept constant from 2009 to 2014. From 2015 onwards it will be subject to an annual degression rate of 5 per cent. The initial tariff is paid for the first 12 years after commissioning, but this can be extended depending upon distance from the coast and water depth. In addition, an 'early bird bonus' tariff is available for wind farms commissioned before 1 January 2016, paid for the duration of the initial tariff payment.[34]

The German Environment Ministry estimates that by the end of 2011, a capacity of around 1500MW could be installed offshore. The German government has a long-term target of up to 25,000MW by 2030. The potential annual electricity yield from offshore wind farms is estimated at 85 to 100TWh by 2030, accordingly. The Infrastructure Planning Acceleration Act, established in 2006, obliges grid operators to ensure grid connection for all offshore plants for which construction has commenced prior to 31 December 2011. This will allow the grid connections of the various offshore wind farms to undergo extensive optimization.[35]

The German government has offered financial support in the form of dedicated funding for the offshore wind sector, such as through non-recourse loans to wind farms, provided by KfW Bankengruppe.[36]

Table E.4 Offshore wind energy German feed-in tariff

Year	Initial Tariff in € cents/kWh	Early Bird Bonus	Basic Tariff in € cents/kWh
2009	13	2	3.5
2010	13	2	3.5
2011	13	2	3.5
2012	13	2	3.5
2013	13	2	3.5
2014	13	2	3.5
2015	12.35	1.90	3.33
2016	11.73	0.0	3.16
2017	11.15	0.0	3.00
2018	10.59	0.0	2.85

Source: German Environment Ministry, 2009

E.5.6 Support mechanisms for development of technologies

The German Environment Ministry has set up a renewable energies research programme, making €50 million available for research and development in the test field.[35]

E.5.7 Other support initiatives

Expert information platforms in the field of offshore wind power have been set up in Bremen, Lower Saxony, Schleswig-Holstein and Mecklenburg-Western Pomerania. They facilitate the exchange of information and experience between plant manufacturers, assembly firms, suppliers, port management companies, shipbuilders and logistical specialists. In the summer of 2005, the German Environment Ministry initiated the creation of an offshore wind energy foundation. It has commissioned construction of a test field of 12 wind turbines for trials.[37]

E.5.8 Deployment

In 2009, Germany installed six offshore turbines, totalling 30MW at one wind farm. This brought the total by the end of 2009 to nine turbines, totalling 42MW across four locations.[38]

The German Environment Ministry estimates that by the end of 2011, a capacity of around 1500MW may be installed offshore, contributing to the German government's long-term target of 25,000MW by 2030.

E.6 Ireland

E.6.1 Offshore wind resource

Ireland experiences one of Europe's highest average wind speeds of approximately 9m/s (see Figure E.7). The offshore wind resource is the highest at the western and northern shores, which are exposed to the prevailing westerly winds. An assessment of the Irish offshore wind resource assuming a water depth below 20m and a minimum distance of 5km from shore gives an estimate of 3300MW, or 11TWh/year.[39]

E.6.2 Wave resource

Ireland experiences one of the most favourable wave regimes in Europe, with a wave energy flux above 60kW/m. The Irish wave energy resource is estimated to be between 187.5TWh[40] and 525TWh.[41] The assessment of the Marine Institute is based on the assumption of a practical wave energy resource possibly exceeding 6000MW and a potential for shoreline and near-shore devices of 800MW.[42]

Figure E.8 illustrates high wave energy fluxes in the north-west, with 70kW/m and 50kW/m for shores facing north and south-west. The Marine Institute quotes a theoretical wave resource of 230TWh/year.[43]

E.6.3 Tidal resource

The theoretical resource for tidal energy in Irish waters has been assessed to be 230,000GWh/year,[43] with the practical resource estimated at 2633GWh/year. Figure E.9 indicates high-velocity tidal currents in the northern and south-eastern waters around Ireland.[44]

E.6.4 Processes for permitting, licensing and allocating seabed rights

The focus of current offshore wind development in Ireland is near shore, within 12 nautical miles of shore. Following introduction of the Foreshore and Dumping at Sea (Amendment) Act 2009, responsibility for certain foreshore functions transferred to the Ministry for the Environment, Heritage and Local Government with effect from 15 January 2010. This includes all foreshore energy-related developments, including wind, wave and tidal energy. The foreshore is the seabed and shore below the line of high water of ordinary or medium tides, and

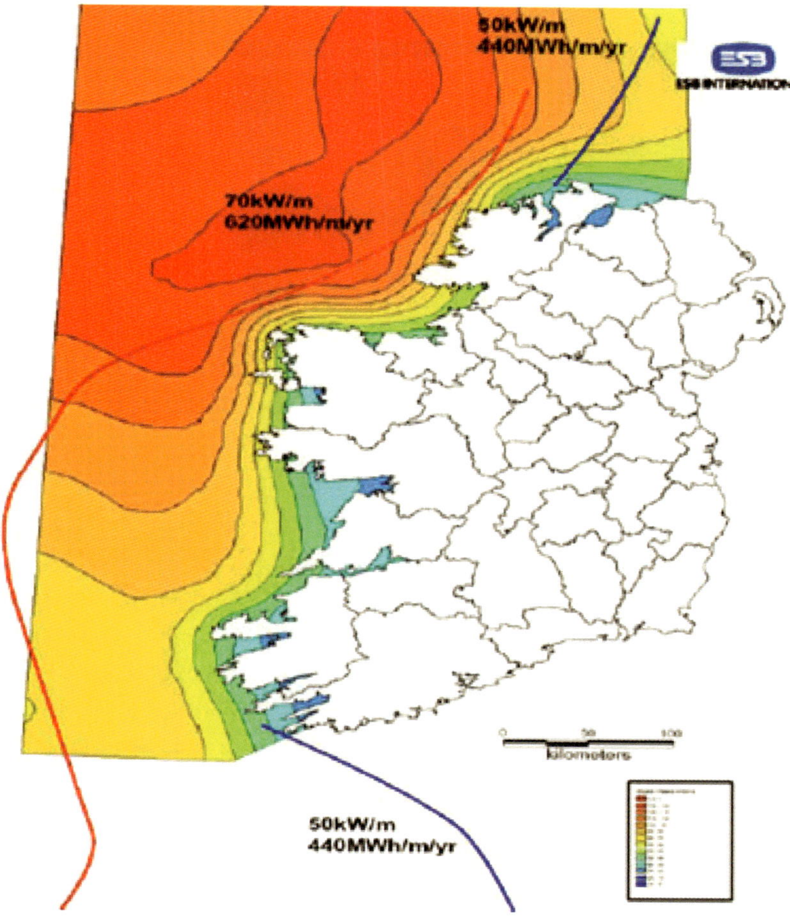

Figure E.8 Indicative wave resource levels in annual MWh/m
Source: ESB International www.esbi.ie

extends outwards to the limit of 12 nautical miles (about 22.24km). The Foreshore Act requires that a lease or licence is obtained for the carrying out of works or placing structures or material on state-owned foreshore. Licences are granted by the minister based on acceptance of an application and payment of fees and at the minister's absolute discretion for acceptance or rejection.[45]

The Department of Communications, Energy and Natural Resources has charged the Ocean Energy Development Unit within the national agency Sustainable Energy Ireland (SEAI), in close collaboration with the Marine Institute in Ireland, to undertake a strategic environmental assessment (SEA) of offshore wind and marine renewable energy in Ireland, which is under way.[46]

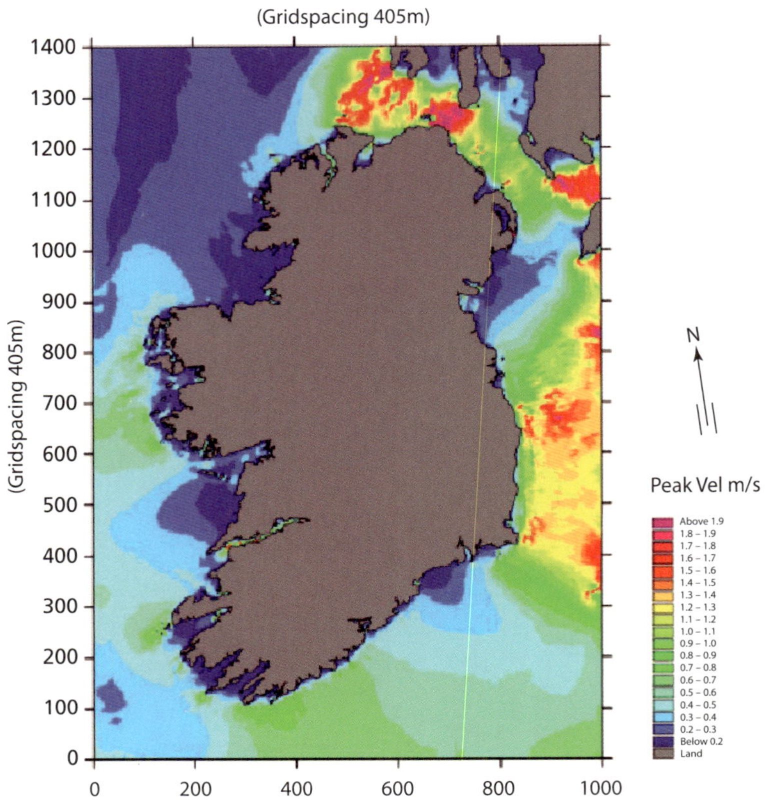

Figure E.9 Depth-averaged peak spring tidal currents
Source: SEAI, Tidal and Current Resources in Ireland

E.6.5 Support mechanisms for project deployment

Since 2006, Ireland has had a renewable energy feed-in tariff (REFIT). The tariffs are theoretically guaranteed for 15 years; in fact, the scheme may not extend beyond 2024. In 2008, the REFIT was amended to provide a FIT of €14 cents/kWh for offshore wind and €22 cents/kWh for ocean energy (wave and tidal).[47] Details on the operating of the proposed REFIT were released in 2009, with further details expected in 2010, following clearance of the scheme with respect to state aid limitations.

The Irish government has set a target of 75MW by 2012 and 500MW of installed ocean energy capacity by 2020.[46]

E.6.6 Support mechanisms for development of technologies

A White Paper released in March 2007 sets out the government's energy policy for 2007 to 2020. It states an intention to support development of offshore wind through a review of cost benefits and research and development (R&D). It also states that the government intends to make Ireland a world leader for ocean energy (OE) technologies within a decade, through the National Ocean Energy Strategy.[47]

Following publication of an Ocean Energy Strategy in 2005, the government announced €26 million of funding to support ocean energy. This led to, among other things, the establishment of the Ocean Energy Development Unit within SEAI and development of the Ocean Energy Prototype Development Fund. The fund aims to stimulate development and deployment of devices and systems. According to SEAI,[46] it focuses on:

- industry-led projects to develop and test wave and tidal energy capture devices and systems;
- independent monitoring of projects/technologies;
- industry-led R&D aimed at the integration of ocean energy within the electricity market and the national electricity grid (and network);
- data monitoring, forecasting, communications and control of OE systems;
- specific industry-led research projects which will be carried out by research centres, third-level institutions and centres of excellence with a high level of expertise in the relevant area.

E.6.7 Deployment

By the end of 2009, Ireland had seven offshore turbines installed, totalling 25.2MW at one site.[48] Ireland's marine energy developments are shown in Table E.5.

Table E.5 Irish marine energy installations

	Stage	Capacity (MW)	Date Commissioned
OE Buoy	Operational	n.c.	2006
Wavebob	Operational	0.35	2007

Note: n.c = not commissioned.
Source: EurObserv'ER (2010)

E.7 Italy

E.7.1 Offshore wind resource

In Italy, mean wind speed at the northern coast is below 5m/s, but can reach up to 7.5m/s in the south. The offshore resource has been estimated to be 3000MW and 10TWh/year, respectively.[49]

E.7.2 Wave resource

Situated in the Mediterranean Sea, Italy experiences a poor wave regime with a wave power flux below 5kW/m, increasing to 10kW/m[50] at favoured sites (e.g. Sicily or Sardinia). Studies assessing the actual power generation potential have not been identified.

E.7.3 Tidal resource

Italy, being located in the Mediterranean Sea, is not exposed to a considerable tidal range. Nevertheless, specific sites such as the Strait of Messina (test location for a 500kW prototype by Fri-El Green Power) might become of interest. No tidal resource studies were available for review.

E.7.4 Processes for permitting, licensing and allocating seabed rights

Publicly available information on permitting, licensing and seabed rights processes in Italy was not readily identified.

E.7.5 Support mechanisms for project deployment

Since January 2002, Italy has had a support mechanism for renewable energy generation based on green certificates. Generators and importers of electricity have minimum obligations for renewable energy generation, which can be met through purchase of the tradable green certificates. Based on the Budget Law 2008, plants that have entered into operation after 31 December 2007 and are larger than 1MW will receive the certificates for 15 years. The certificate price is equal to the difference between a reference value (€18 cents/kWh in 2008) and the annual average electricity sale price. Plants smaller than 1MW capacity have the option of choosing either the green certificates or to receive a feed-in tariff for 15 years instead. The tariff for wave and tidal is €34 cents/kWh.[51]

Investments in renewable energy in the southern regions of Italy also benefit from a rebate on value-added tax (VAT) (10 instead of 20 per cent) and from a ten-year corporate tax reduction.[52]

E.7.6 Support mechanisms for development of technologies

Italian public funding for device development is available under the Industria 2015 programme, announced by the Ministry of Economic Development. One recipient is Project GEOMA, a project to develop a hybrid concrete/steel 3.5MW floating wind turbine intended for deep-water applications, such as in the Mediterranean Sea.[53]

E.7.7 Deployment

Italy currently has no offshore wind turbines installed. It has one marine device operational, the 20kW Kobold System.[54]

E.8 Japan

E.8.1 Offshore wind resource

Figure E.10 presents a wind resource map of Japan from a meteorological numerical model.[55] The model shows that onshore winds are mostly below 5m/s, while coastal and offshore wind speeds vary depending upon geographic orientation of the coast between 5.5m/s and 8m/s. Offshore development faces the challenge of deep waters, particularly at the east coast of the country, which experiences higher wind speeds than the west coast. On the east coast, water depths of 100m and above are common at only 10km distance from shore. In most locations, water depth increases rapidly. Nonetheless, Japan's government deems 3000MW of wind capacity a feasible near-term target in its Primary Energy Supply Plan. Meanwhile, the Japanese Wind Power Association proposes a wind capacity of 11,800MW by 2030.[56]

E.8.2 Wave resource

Japan experiences a low to moderate resource with a wave energy flux between 10 and 20kW/m.[57] Studies assessing the actual power generation potential have not been identified.

E.8.3 Tidal resource

No tidal resource assessments were available.

E.8.4 Processes for permitting, licensing and allocating seabed rights

Publicly available information on permitting, licensing and seabed rights processes in Japan was not readily identified.

E.8.5 Support mechanisms for project deployment

Japan has a renewable portfolio standard (RPS) and a FIT, but the latter only for solar photovoltaics (PV).[58]

E.8.6 Support mechanisms for development of technologies

The New Energy and Industrial Technology Development Organization (NEDO) is responsible for R&D activities involving renewable energy in Japan. It actively

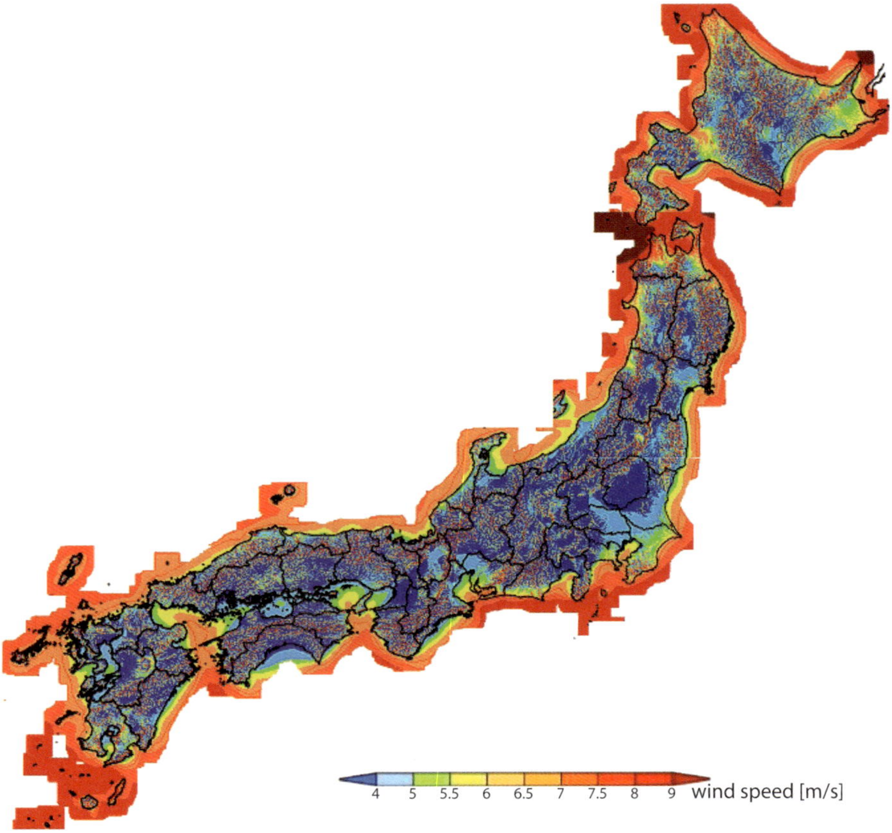

Figure E.10 Wind Resource Map, Japan
Source: Takagi et al, 2006

undertakes the development of new energy and energy conservation technologies, verification of technical results, and introduction/dissemination of new technologies (e.g. through support for introduction). From 2008 to 2013, a 200 million yen (€1.5 million) per year project into offshore wind aims to identify ocean wind characteristics, and meteorological and oceanographic conditions specific to Japan, in order to develop suitable technologies for ocean wind observation and wind power generation, and to establish an appropriate environmental impact assessment approach. It aims to identify and utilize suitable offshore research sites.[59]

The Japan Agency for Marine-Earth Science and Technology (JAMSTEC) supports the development of marine technologies.[60]

E.8.7 Deployment

Japan has two offshore wind turbines (600kW each) installed off the coast of Setana in Hokkaido.[61]

During the early 1990s, an oscillating water column (OWC) prototype 60kW converter was integrated within a breakwater at the port of Sakata, Japan.[62] From 1978 to 1986, JAMSTEC tested a large-scale floating-wave generating equipment, 'Kaimei' (80m long, 12m wide), partly as joint research with the International Energy Agency (IEA). Subsequently, an offshore floating-wave generator prototype, 'Mighty Whale' (110 kW, 50m long, 30m wide), was constructed and tested offshore from 1998.[59]

E.9 The Netherlands

E.9.1 Offshore wind resource

The Netherlands are exposed to wind speeds above 8m/s (see Figure E.7), indicating a high offshore wind resource which is estimated to be around 10,000MW, or 33TWh/year.[63]

E.9.2 Wave resource

With the coast facing in north-westerly directions and being sheltered by the British Isles, the Dutch coast experiences a moderate wave regime with an average annual wave flux below 10kW/m.[64] Studies assessing the actual power generation potential have not been identified.

E.9.3 Tidal resource

No relevant studies on the Dutch tidal resource were available for this book.

E.9.4 Processes for permitting, licensing and allocating seabed rights

The Dutch government's 2008 draft National Water Plan looks at North Sea offshore wind power potential, alongside interests in fisheries, shipping, sand extraction, oil and gas extraction, nature conservation and coastal defences. It proposes two areas for wind energy deployment, which together could theoretically accommodate over 9000MW of wind power, assessing grid access options and related transmission costs. The final plan (pending as of February 2010) will set the areas in which permits for offshore wind farms will be made available.[65]

E.9.5 Support mechanisms for project deployment

The Netherlands reintroduced a subsidy scheme in 2007 after the previous scheme was stopped in 2006. Known as the *Subsidies Duurzame Energie* (SDE), it pays a premium above the market rate. In February 2009, a base rate of €18.6 cents/kWh was announced for offshore wind, paid to projects for 15 years.

There is a subsidy ceiling of €2645 million in total, after which no more capacity will be supported. It is expected to provide support for 450MW of offshore wind capacity. A subsidy for marine energy is not specified.[66]

The Netherlands also has an Energy Investment Deduction Scheme which grants tax incentives for investment in renewable energy projects.[66]

E.9.6 Deployment

By the end of 2009, the Netherlands had 130 offshore turbines installed, totalling 246.8MW across four wind farms.[67] It has one marine energy device under development, the 45kW Tocardo.[68]

E.10 Norway

E.10.1 Offshore wind resource

Norway has a long exposed shoreline facing the North Sea, the Atlantic and the Bering Sea, with offshore constructions facing the challenge of deep waters relatively close to shore. In 2008, the Norwegian water and energy department (Norges vassdrags- og energiedirektorat) assessed the considerable Norwegian offshore wind resource. At 50m above ground, mean wind speeds for the western coast of Norway are between 7m/s and 9m/s. Potential locations for offshore wind developments were considered according to water depth (10m, 20m and 100m) and a distance from shore (1km up to 10km). The results showed that the offshore potential was highly influenced by these factors as the potential varies between the following figures:

- around 6000MW: water depth of up to 20m and minimum distance from shore of 10km;
- around 30,000MW: water depth of 20m and minimum distance from shore of 1km; and
- around 140,000MW: for a maximum water depth of 100m with a maximum distance of 20km.[69]

Figure E.11 provides an overview of wind speeds experienced 80m above ground. NVE engaged Kjeller Vindteknikk to produce a new wind map covering the Norwegian landmass and the surrounding sea.[70] The total theoretical potential for Norway's offshore wind resource has been estimated to be around 14,000TWh/year.[71]

E.10.2 Wave resource

Norway experiences a very good wave regime, with variations regarding geographical and seasonal conditions. The south of Norway's coast is sheltered

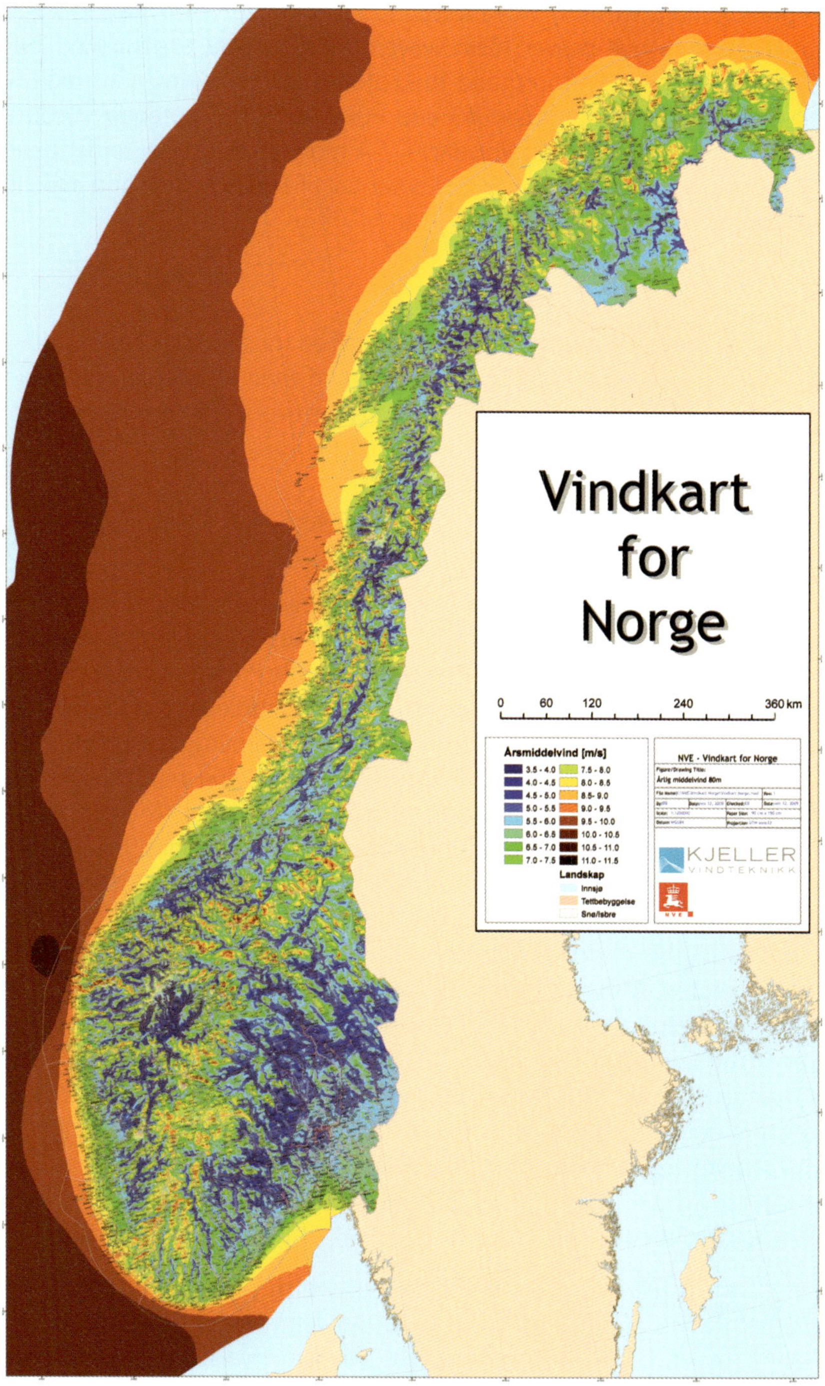

Figure E.11 Wind Resource Map, Norway
Source: Kjeller Vindteknikk AS, 2009

from predominant westerly wave direction reaching between 10 and 25kW/m annual average wave power. Best wave conditions are experienced further north between the Sognefjord and the Lofoten Islands. Annual average wave power reaches figures between 40 and 60kW/m.[72] The potential resource is estimated to be around 400TWh/year.[73, 74] A more recent study undertaken on behalf of Enova suggests a theoretical potential of 600TWh/year and a technical or realistic potential of 12 to 30TWh/year.[75]

E.10.3 Tidal resource

The Norwegian tidal resource has been quantified for specific locations north of Bodø. Twenty-four locations were assessed and their individual potential was found to be between 3 and 758GWh/year, respectively. The overall resource has been estimated to be around 1TWh/year.[75]

E.10.4 Processes for permitting, licensing and allocating seabed rights

A governmental proposal for a new offshore renewable energy act and national strategy was scheduled for hearing in the Norwegian parliament before the summer of 2010. According to the Norwegian Trade Portal, the new concession policy resembles the licensing rounds for offshore petroleum blocks, with public administration and control of the management of the energy resources offshore.

Offshore areas will be opened for applications following impact assessments, for early consideration and balancing of relevant interests. The proposed act includes regulations on the process of applying for concessions, establishment, operation and close-down of offshore renewable energy production and the offshore grid. Regulations on compensation to fishermen, similar to the regulations in the petroleum sector, are proposed, as well as coverage of concerns related to security and working environment, area fees, system operation and export and import of electrical energy

Under the Energy Act of 1990, the Norges vassdrags- og energidirektorat (NVE) can grant licences to wind farms within the sea boundary (i.e. located in the waters between the mainland and nearby islands). Projects bigger than 10MW require an environmental impact assessment.[76]

E.10.5 Support mechanisms for project deployment

The Norwegian government's Energy Fund, which provides up to 25 billion Norwegian krone (~ €3.3 billion) over a ten-year period, is administered by Enova.[75] Enova is a public enterprise, owned by the Royal Norwegian Ministry of Petroleum and Energy, which has operated a programme to provide investment support for ocean energy demonstration projects since 2009. Examples of projects awarded support include tidal, wave and offshore wind demonstrators, including onshore qualification testing of a 10MW offshore wind turbine.[77]

E.10.6 Support mechanisms for development of technologies

Eight Centres for Environment-Friendly Energy Research (CEERs) have been established through the Research Council of Norway to pursue renewable energy and energy efficiency research. Each CEER receives up to 20 million Norwegian krone (€2.3 million) over the next five years, in addition to private finance from industry partners. Two of the centres are focussed on offshore wind. The Norwegian Centre for Offshore Wind Energy (NORCOWE) is an interdisciplinary resource centre which aims to take the lead in the development of new innovative and cost-efficient solutions for the exploitation of offshore wind energy at large water depths and in harsh offshore environments.[78] The Norwegian Research Centre for Offshore Wind Technology (NOWITECH) focuses on deep-sea (> 30m) technologies, including bottom-fixed and floating turbines. Its work packages cover integrated numerical design tools; energy conversion systems; novel substructure; grid connection and system integration; operation and maintenance; and assessment of novel turbine concepts.[79]

In addition, the Research Council of Norway operates a thematic Research Trade and Development (RTD) programme, RENERGI, providing funding for energy-related projects not covered by the eight centres.

E.10.7 Deployment

In 2009, Norway installed one offshore turbine of 2.3MW. This is currently its only offshore wind installation.[80]

A Norwegian government-funded R&D project led to the construction in 1985 of two full-sized (350kW and 500kW rated power) shoreline prototypes near Bergen.[81]

A wave energy converter concept, Seabased, is being tested offshore outside Runde, an island on the west coast of Norway. The two test units are in the order of 20kW. The project represents a cooperative venture between Vattenfall and Tussa Energi.

According to the Norwegian Trade Portal, an increasing number of applications are being made for offshore wind projects in Norway, both fixed sea bottom projects in shallow waters and fully floating foundation projects for deeper seas. There are currently 19 notifications for offshore wind parks in Norway, 7 of which are floating installations. For example, starting in 2010, StatoilHydro's Hywind was to place a 2.3MW floating turbine 10km off the coast north of Stavanger for a two-year test period. Another floating turbine company, Sway, also plans to test a turbine for two years. [82] Vestavind Kraft plans to test a floating wind turbine in three stages offshore Selje and Vågsøy.

All three of these are relatively small pilot phase projects only. Licensing policy issues are causing delay to some similar projects.[83]

Other wave and tidal demonstration projects have also been provided state funding. An example is the deployment by Hydra Tidal during 2010 of a tidal power prototype in the Gimsøystraumen tidal current at Lofoten Islands, North Norway.

E.11 UK

E.11.1 Offshore wind resource

Situated in an exposed location between the Atlantic Ocean and the North Sea, the UK has a high wind energy resource. The resource is estimated to be around 70GW, translating into an annual output between 230 and 334TWh.[84] Figure E.12 provides an overview of the distribution of the offshore wind resource in UK waters at 100m above water.[85]

E.11.2 Wave resource

Wave power varies greatly along the coasts of the UK. The south and east coasts experience a low to moderate wave regime with a range below 5 up to 20kW/m. Off the north-west coast of Scotland, wave power levels are typically around 60 to 70kW/m. Figure E.13 provides an overview of the distribution of the wave resource.

Thorpe estimated the overall resource at 120GW.[86] The Carbon Trust suggests a practical offshore wave energy resource to be around 50TWh/year.[87] The UK practical near-shore and shoreline wave energy resources have been re-estimated at 7.8TWh/year and 0.2TWh/year, respectively.[88]

E.11.3 Tidal resource

The distribution of the tidal energy around the UK is indicated in the BERR *Atlas of UK Marine Renewable Energy Resource* and shown in Figure E.14.

In 2005, Black & Veatch assed the total UK tidal stream resource to be around 110TWh/year. Further analysis suggested a figure of 18TWh/year for the technically extractable resource, representing 10 to 15 per cent of the known worldwide tidal stream resource. Only around 20 per cent of the identified tidal stream sites are deemed to be suited to near-term developments as they are located in moderate water depth between 30m and 40m and have a tidal stream velocity between 2.5m/s and 4.5m/s.[89] Black & Veatch have recently completed an updated UK tidal resource assessment which estimates a significant increase from the 18TWh/year figure. The report is to be published in early 2011.[90]

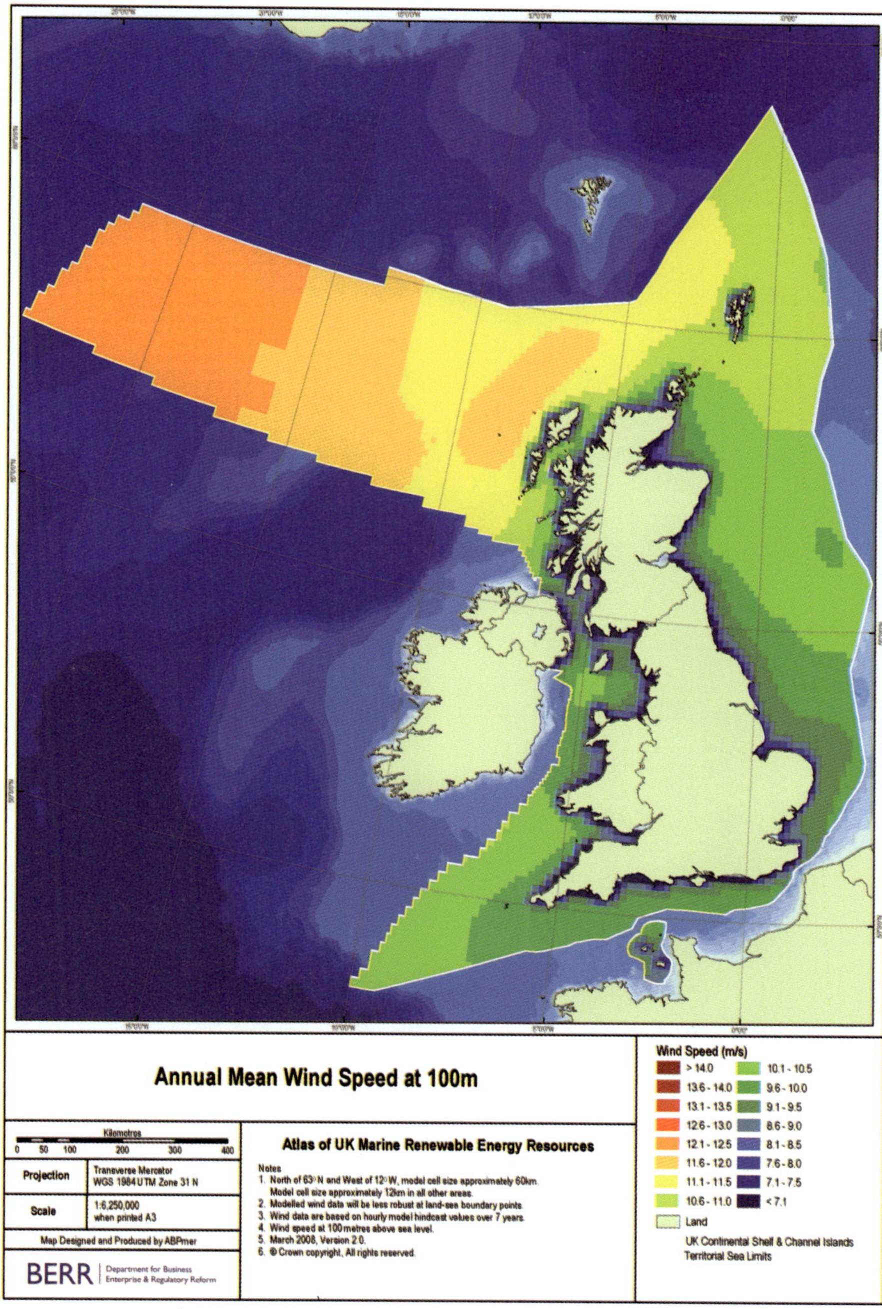

Figure E.12 UK annual mean wind speed at 100m

Source: BERR, Atlas of UK Marine Renewable Energy Resource, 2008 (Crown copyright)

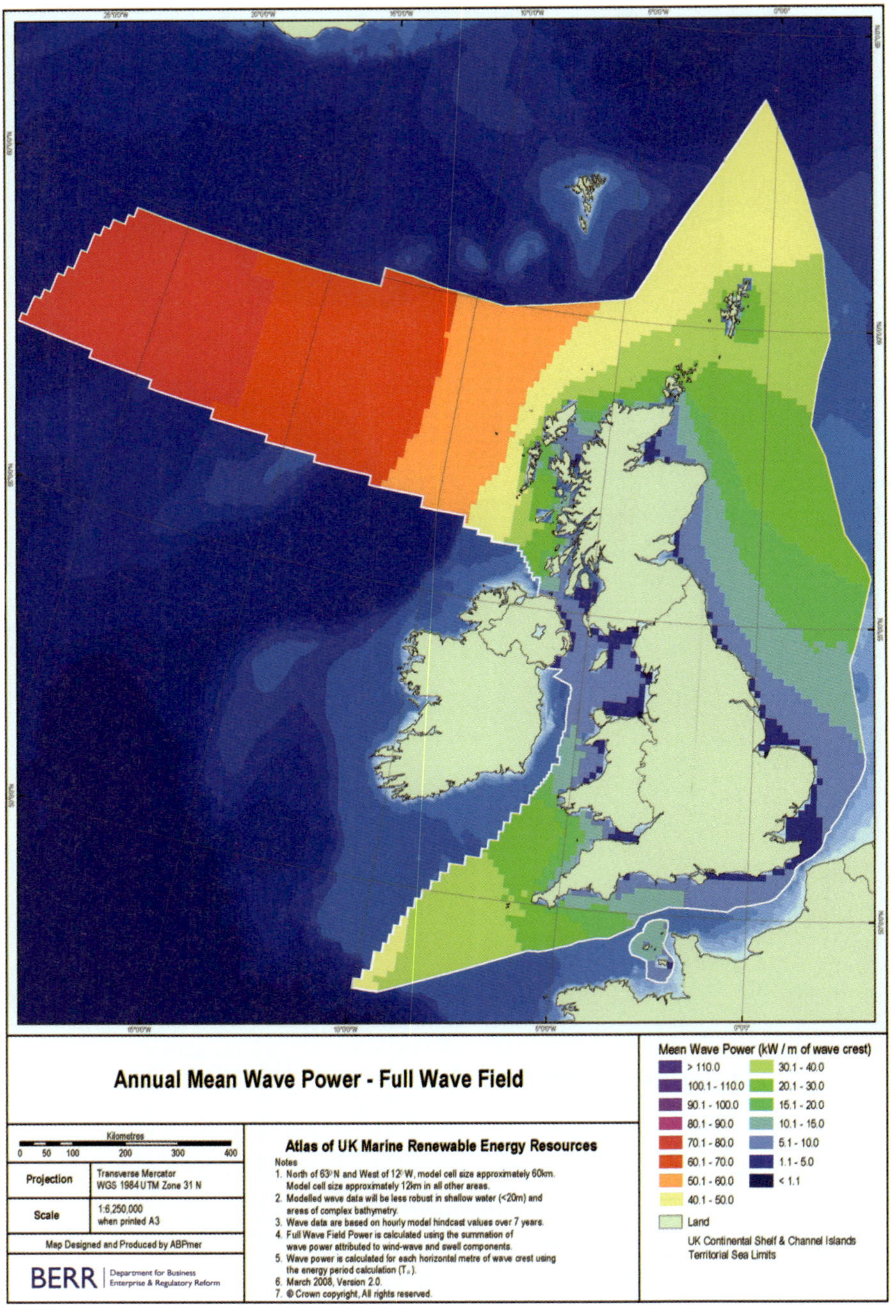

Figure E.13 UK annual mean wave power: Full wave field

Source: BERR, Atlas of UK Marine Renewable Energy Resource, 2008 (Crown copyright)

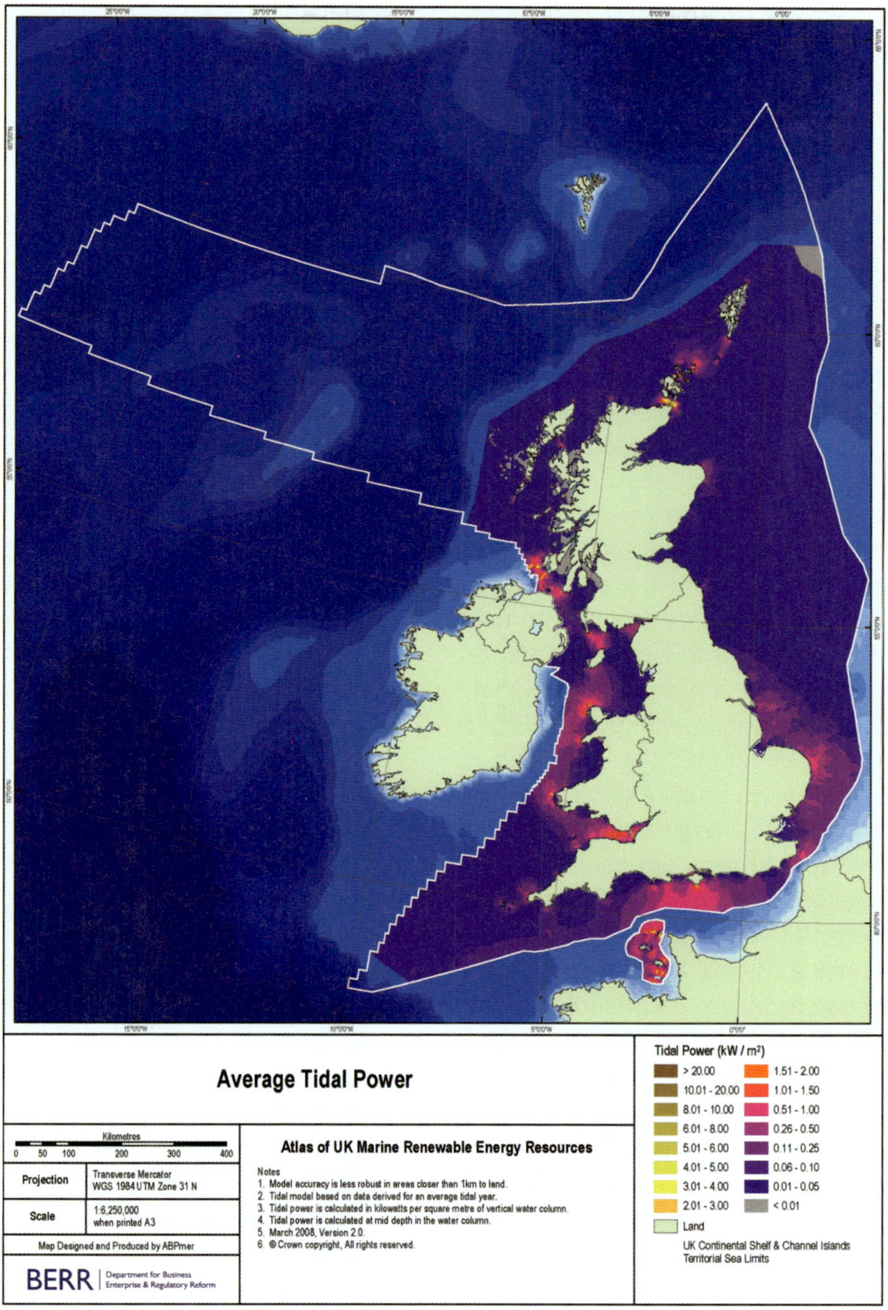

Figure E.14 UK average tidal power

Source: BERR, Atlas of UK Marine Renewable Energy Resource, 2008 (Crown copyright)

Sites particularly suited for tidal stream developments in the UK are identified to include the following:

- Pentland Skerries with ~3900GWh/year;
- Storma Pentland Firth with ~2800GWh/year;
- Duncasby Head with ~2000GWh/year; and
- Casquets with ~1700GWh/year.[91]

E.11.4 Processes for permitting, licensing and allocating seabed rights

The UK seabed is the property of the Crown Estate. Offshore wind development in the UK has had three rounds. Each round starts by the Crown Estate setting aside zones for proposed wind development, based on the government's offshore energy strategic environmental assessment (SEA) that identifies areas where there are no overriding environmental concerns. Companies are then invited to bid for sites, and once awarded, lease holders can submit planning applications. Round 1 and 2 sites, awarded in 2000 and 2003, are now mostly built or under construction, with the total installed capacity expected to reach 8GW. The round 3 development process was launched in June 2008.[92] In January 2010, it was announced that round 3 rights have been granted in nine UK coastal zones, with potential for 6400 additional turbines and a combined capacity of 32GW.[93]

The Crown Estate also has the power to grant leases for areas of the seabed to offshore energy developers. Securing a site lease requires the project developer to carry out comprehensive environmental impact assessments and monitoring, as well as demonstrating appropriate project design. Following successful issue of a seabed lease, developers must obtain relevant consents from the Marine and Fisheries Agency (in England and Wales), the Fisheries Research Service (Scotland) or the Northern Ireland Environment Agency, including for drilling and foundation work and for establishing an electrical connection to shore. The process necessitates consultation with a wide variety of bodies' needs and generally takes at least six months.[94]

The Planning Act 2008 of November 2008 aims to provide a streamlined consenting process for nationally significant infrastructure projects in England and Wales, including certain types of energy. The number of applications and permits required for such projects will be reduced compared with the current legislation. The act also allows for the Infrastructure Planning Commission (IPC) to be established. This is an independent body responsible for examining development consent applications for nationally significant infrastructure projects, including offshore energy projects over 100MW in England and Wales. It aims to resolve any potential issues in advance of submission of the planning application and to produce consent decisions in nine months.[95]

Offshore energy projects of less than 100MW capacity will be licensed by the Marine Management Organization (MMO), a public body created through the Marine Bill. The Marine Bill also sets the framework for a new offshore planning system, based on long-term objectives for the UK's marine area. The new system aims to reduce the planning process to just one licence for offshore energy developments.[96]

E.11.5 Support mechanisms for project deployment

The UK government published a *Renewable Energy Strategy* in July 2009 which proposes 14GW of offshore wind by 2020 (see Figure E.15). The UKERC Marine Renewable Technology Roadmap sets a target of 2GW of wave and tidal energy installed by 2020 in UK waters.[97] The Crown Estate has set a target for the Pentland Firth strategic area of 700MW of offshore wave and tidal stream generation by 2020.[98]

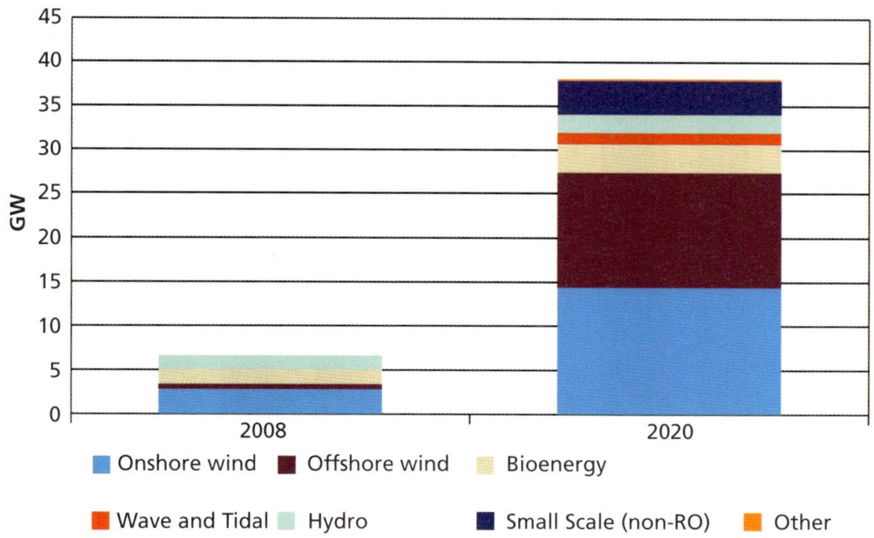

Figure E.15 Renewable electricity technologies: Actual capacity in 2008 and projected capacity in 2020 for UK
Source: HM Government (2009)

Financial support for renewable energy generation is provided through the Renewables Obligation (RO) and Renewables Obligation (Scotland) (ROS) and Renewables Obligation (Northern Ireland) (NIRO). Banding was introduced in April 2009 so that two renewable obligation certificates (ROCs) are granted for every MWh of electricity generated from a marine energy device. This is over £90/MWh (€10 cent/kWh) and means that marine energy generators will receive approximately £12.8 pence/kWh (€14 cent/kWh) when the income received from the electricity generated, levy exemption certificates and the recycle of ROCs is also included. Following approval from the European Commission, the Scottish government plans to increase the support available through the ROS to three ROCs per MWh of electricity generated from tidal stream devices and five ROCs per MWh of electricity generated from wave energy devices.[98] After initially being set at 1.5 ROCs/MWh, the government reviewed the RO banding for offshore wind and announced that all offshore wind farms gaining full accreditation between April 2010 and March 2014 will receive 2 ROCs/MWh.[99] Historic ROC prices are shown in Figure E.16 for the period October 2002 (start of CP1) to January 2010 (mid CP8). The price has historically fluctuated around £48/MWh (~ €5.5 cents/kWh).[100]

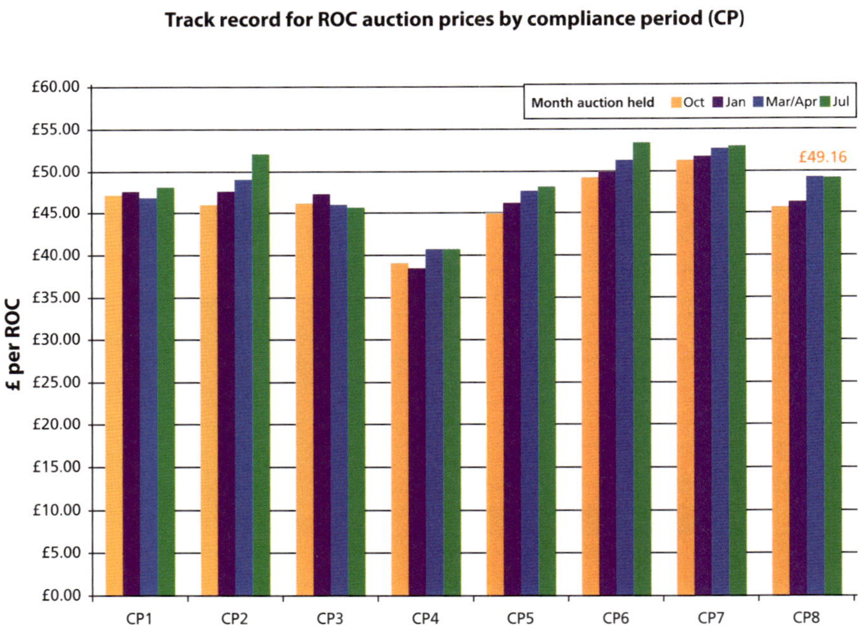

Figure E.16 UK historic renewable obligation certificate (ROC) prices
Source: NFPA, 2010 (Non-Fossil Purchasing Agency)

Capital grant schemes include the UK government's Marine Renewable Deployment Fund (MRDF) (25 per cent capital grant for qualifying projects and an enhanced payment of £10 pence/kWh (€11 cent/kWh) in addition to the normal ROC payment). It was established by the UK government in 2004 to support the first devices operating at sea. UK£42 million of the £50 million MRDF (€46.2 million of €55 million) was allocated to provision of a combination of capital grant and revenue support for devices. Although a small number of applications for the MRDF have been received, to date no projects have succeeded in fulfilling the eligibility criteria of three months full-scale device sea trial data. The Marine Renewables Proving Fund (MRPF) is a new fund of up to £22 million (€24 million) launched by the government in the *Renewable Energy Strategy* of 2009. In Scotland, the Scottish ministers' Wave and Tidal Energy Support Scheme (40 per cent capital grant for qualifying projects and an enhanced payment of £10 pence/kWh in addition to the normal ROC payment), launched in 2006 and worth UK£13.5 million (€15 million) in support, has provided support to nine schemes so far.[101]

E.11.6 Support mechanisms for development of technologies

The direct total government spending on technology R&D (provided through the UK Department of Trade and Industry (DTI) New and Renewable Energy Programme) for 1999 to 2005 was UK£3.9 million (€4.3 million) for wave energy and UK£6.5 million (€7.2 million) for tidal stream energy. During the same period, the Research Council spend on its wave and tidal research programme was UK£3.6 million (€4 million). As well as the government's capital grant schemes (see above), grants are provided by a number of different bodies (e.g. the Carbon Trust's Marine Energy Accelerator and Energy Technology Institute (ETI) Marine Energy Programme).[102]

The Saltire Prize is a UK£10 million (€11 million) prize challenge for advances in wave and tidal energy launched by the Scottish government in December 2008. It is to be awarded to the team that can demonstrate in Scottish waters a commercially viable wave or tidal energy technology that achieves a minimum electrical output of 100GWh over a continuous two-year period using only the power of the sea and is judged to be the best overall technology after consideration of cost, environmental sustainability and safety.[103]

E.11.7 Other support initiatives

The Crown Estate is funding a UK£5.15 million (€5.7 million) research and development project looking at solutions to the problem of radar interference and wind turbines. The 19-month R&D programme examines how the impacts of wind turbines upon the NATS En Route primary radar infrastructure can be mitigated, working with Raytheon Canada, the suppliers of the NATS systems. UK£1.6 million of the funding comes from wind companies, UK£2 million from The Crown Estate and UK£1.55 million from the UK Department of Energy and Climate Change (DECC). The Crown Estate has also funded seabed surveys and aerial bird surveys in order to support offshore wind developments.[104]

E.11.8 Deployment

In 2008, the UK overtook Denmark as the country with the highest operating offshore wind capacity in the world (UKERC, 2009). In 2009, the UK installed 84 offshore turbines, totalling 284.4MW across 3 separate wind farms. It now has 287 turbines, totalling 882.8MW across 12 sites.[105]

By the end of April 2009, the UK had 0.5MW installed capacity of wave energy and 1.45MW of tidal stream energy.[102] The installed devices are shown in Table E.6.

Table E.6 UK marine energy installations

	Stage	Capacity (MW)	Date Commissioned
Limpet	Operational	0.5	2000
Seaflow	Operational	0.3	2003
Open Centre Turbine	Operational	0.25	2008
SeaGen	Operational	1.2	2008
Pulse Hydrofoil	Being tested	0.1	2009
Sea Snail	Being tested	0.15	n.c.

Note: n.c = not commissioned.
Source: EurObserv'ER (2010)

Further projects are also currently under development and are expected to be installed in the next few years. For example, early in 2008, Marine Current Turbine (MCT), developer of the SeaGen turbine, joined in partnership with npower renewables to develop a tidal stream project under the management of a newly created company, SeaGen Wales. The plan is for a 10.5MW farm to be located in The Skerries, off the north-west coast of Anglesey. Completion is expected in 2011/2012. Another venture for MCT, announced in November 2008, is a tidal scheme for the Pentland Firth, Scotland. The company will apply for a lease from the Crown Estate and, subject to finance and gaining consent, plans a 50MW plant by 2015.[106]

Pelamis Wave Power's 0.75MW was previously installed at EMEC for testing, but was removed after a successful trial period. Pelamis Wave Power is now developing the next generation Pelamis device, 'P2' (0.75MW), which will be installed and tested at EMEC during 2010.[107] Projections by BWEA/Entec for growth of UK marine energy capacity are shown in Figure E.17. Meeting these targets will mainly be linked to the delivery of up to 1.2GW of wave and tidal projects which have been granted leases by the Crown Estate in the Pentland Firth and Orkney area in March 2010.[108]

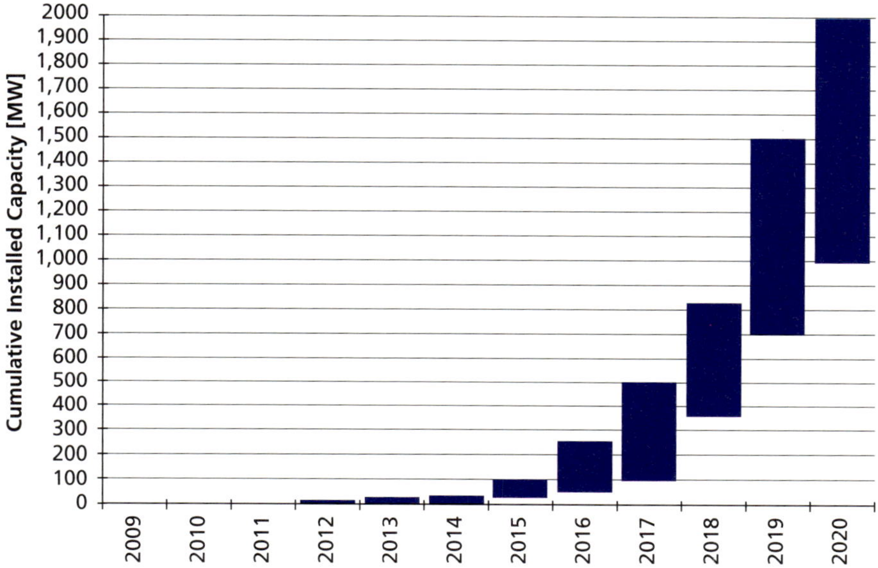

Figure E.17 Potential UK cumulative installed capacity of marine energy projects to 2020
Source: BWEA/Entec 2009

E.12 Belgium

E.12.1 Offshore wind resource

In 2004, the offshore wind energy potential on the Belgian continental shelf was assessed to be between 2GW and 4GW.[109] A review of offshore wind energy undertaken by a consortium led by the University of Delft gives a resource estimate of 1.2MW and 4TWh/year, respectively.[110]

E.12.2 Wave resource

Due to the sheltered location at the Channel and the southern North Sea, the wave energy flux is around 5 to 10kW/m, rendering wave energy an unattractive option.[111]

E.12.3 Tidal resource

While Belgium experiences tidal fluxes, no reference to tidal resource assessments has been found.

E.12.4 Processes for permitting, licensing and allocating seabed rights

Publicly available information on permitting, licensing and seabed rights processes in Belgium was not readily identified.

E.12.5 Support mechanisms for project deployment

National targets and policies for renewable energy in Belgium are regional rather than national, split between Wallonia, Flanders and Brussels. All three regions have established quota systems. The certificates have historically been around €109/MWh (Flanders) and €90/MWh (Wallonia). For offshore wind, there is currently no market price; but a federal minimum price has been set. For the first 256MW installed, this is €10.7 cents/kWh for the first ten years of operation and €9 cents/kWh for the second ten years of operation. For additional installed capacity beyond the initial 256MW, a minimum tariff of €9 cents/kWh applies for a 20-year period.[112]

E.12.6 Support mechanisms for development of technologies

Investment support schemes are available for renewable energy technologies.

E.12.7 Deployment

Belgium has six offshore wind turbines, totalling 30MW at one site.[113]

E.13 Finland

E.13.1 Offshore wind resource

The Finnish offshore wind energy resource is estimated to be around 6000MW and around 20TWh/year, respectively.[114]

E.13.2 Wave resource

Finland has shores at the Baltic Sea experiencing wave fluxes below 5kW/m and possibly up to 10kW/m in the south-western region of Hanko.[115]

E.13.3 Tidal resource

Due to the geography of the Baltic Sea, the coastal regions experience only very low tides, around 10cm in height difference.

E.13.4 Processes for permitting, licensing and allocating seabed rights

The Electricity Market Act (386/1995) provides guaranteed grid access for all electricity generators and users, including renewable energy generators.[116]

E.13.5 Support mechanisms for project deployment

Energy Aid is a state grant provided for investments and research into renewable energy. The preparation and planning costs and the costs of materials are eligible for subsidies, amongst other costs.[117] The construction costs of renewable energy plant are co-financed by the government with subsidies (e.g. up to 40 per cent in the case of wind). The government has imposed a tax per kWh on electricity suppliers. This is then provided to renewable generators at €0.69 cents/kWh for wind and €0.42 cents/kWh for other eligible technologies.[116]

E.13.6 Support mechanisms for development of technologies

Publicly available information on support for development of technologies in Finland was not readily identified.

E.13.7 Deployment

Finland has eight offshore wind turbines, totalling 24MW at one site.[118]

E.14 Spain

E.14.1 Offshore wind resource

The Spanish offshore wind resource is estimated to be around 2000MW, or 7TWh/year.[119]

E.14.2 Wave resource

The Mediterranean shores of Spain feature a very low energy flux, with below 5kW/m. North and west Spain shores are exposed to the Atlantic Ocean and experience a good wave climate, with intensities between 30 and 60kW/m.[120]

E.14.3 Tidal resource

To date, no comprehensive assessment of the tidal resource at the Spanish coast has been undertaken. The Ria de Muros, an estuary on the Galician coast, has been assessed based on a numerical model.[121] The estuary has a tidal range of 2.5m and a current velocity of about 2m/s. Further research would be beneficial.

E.14.4 Processes for permitting, licensing and allocating seabed rights

The approval of offshore wind farms is regulated by Royal Decree 1028/2007 of 20 July 2007, which establishes a procedure for competition of offshore wind power exceeding 50MW.

Developers may apply to establish an offshore wind farm within areas identified by the government's *Marine Wind Map*, shown in Figure E.18, as suitable (with or without constraints). At this point, rival developers may also bid for the specific area, submitting technical documentation and a tariff offer to the governmental review committee. The selected developer is eligible to apply for full administrative approval within the two years after selection, during which time they may conduct necessary studies.

An application for full administrative approval is submitted to the Directorate General for Energy Policy and Mines, Department of Sustainability and the Sea Coast and the Department of Environmental Quality and Assessment, the latter with respect to approval of the project EIA specifically. The application is then licensed following standard procedures for electric power facilities, according to Royal Decree 1955/2000 of 1 December 2000.[122]

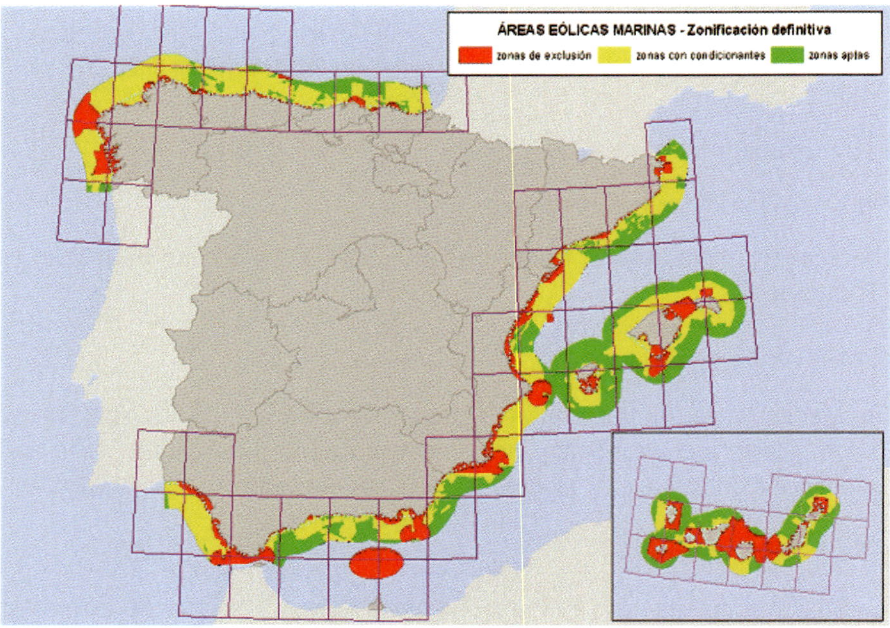

Figure E.18 Spanish Marine Wind Map: Allowable areas for offshore wind farm development

Source: Spanish Ministry of Environment and Rural and Marine Areas (2009)

E.14.5 Support mechanisms for project deployment

For renewable energy plants between 50MW and 100MW capacity, a bonus is paid above the market price received for electricity generation. Plants up to 50MW can choose between the bonus payment and a guaranteed feed-in tariff. Offshore wind is eligible, up to a combined cap of 20,155MW across onshore and offshore wind. However, it is only eligible for plants up to 100MW. Wave energy, tidal energy, ocean thermal energy and ocean current energy are also eligible and there is no cap in terms of total capacity to be supported for these technologies. The FIT is received for the entire operation of a system, but is reduced after a certain period of operation: 25 years for marine energy and 20 years for wind energy.[123]

For wind farms operational since January 2008, the fixed tariff option is €7.32 cents/kWh, reduced to €6.12 cents/kWh after 20 years of operation. The fixed bonus option is €2.93 cents/kWh.[124]

In the Basque Country there is a wave energy target of 5MW.[125]

E.14.6 Support mechanisms for development of technologies

A €30 million initiative was announced in January 2010 to support offshore energy research in Spain. A consortium of 19 companies and 25 research centres will undertake the three-year Ocean Lider Initiative. It is backed by Spain's Ministry of Science and Innovation and has received grants from the Spanish Centre for the Development of Industrial Technology (CDTI) and the State Fund for Local Investment. It claims to be the world's largest marine energy R&D project. The initiative aims to develop integrated offshore energy installations, including by combining them with proven energy sources such as offshore wind power.[126]

E.14.7 Deployment

Ocean Power Technologies (OPT) is developing a 1.39MW project in the north of Spain. A 40kW PowerBuoy device was deployed there in 2008.[127] A demonstration project, Mutriku, of 300 kW is under construction and due for completion in 2010. An R&D facility, bimep, that can host 20MW is at the licensing and design stage and aims to be ready in 2011.[128]

E.15 Sweden

E.15.1 Offshore wind resource

Sweden experiences a good wind climate, with the offshore wind resource estimated at around 7000MW, or 22.5TWh/year, considering sites at water depth below 40m and a maximum distance from shore of 30km.[129]

E.15.2 Wave resource

Most of Sweden's shoreline lies in the sheltered Baltic Sea, with wave fluxes below 5kW/m. The North Sea coast is located in a sheltered position close to Denmark.[130]

E.15.3 Tidal resource

Sweden's coast is mainly enclosed by the Baltic Sea, which experiences only a small (~10cm) tidal variation. Research referring to tidal stream resource has not been found.

E.15.4 Processes for permitting, licensing and allocating seabed rights

Publicly available information on permitting, licensing and seabed rights processes in Sweden was not readily identified.

E.15.5 Support mechanisms for project deployment

Sweden has an electricity certificate mechanism for the promotion of renewables. It is un-banded, with all technologies receiving one certificate per MWh generated. Wind and wave energy generation is eligible under the system. The certificates are received for 15 years or until the end of 2030, whichever is earlier. No wave energy plants are currently registered under the scheme. The average price of the certificates from 2003 to 2009 is shown in Figure E.19, equating to a range of approximately €1.5 to 4.2 cents/kWh.[131]

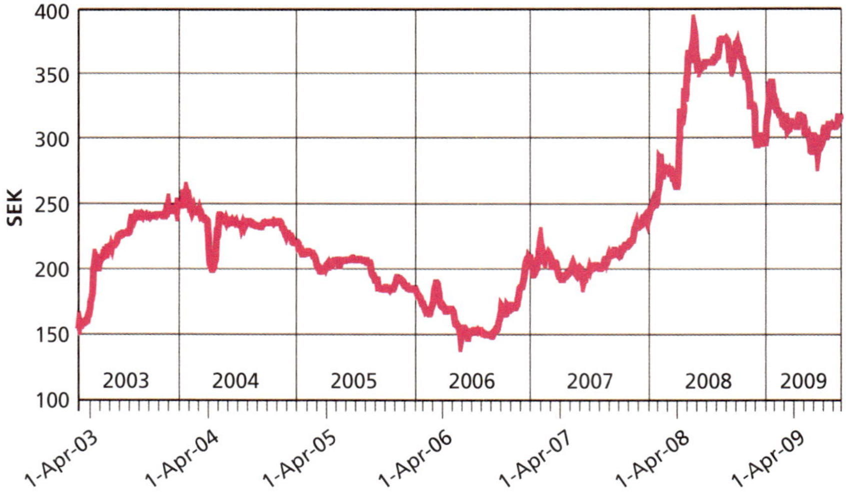

Figure E.19 Average price of spot-traded electricity certificates
Source: Swedish Energy Agency (SEA)

In 2008, a Swedish government bill named Coordinated Climate and Energy Policy – Energy proposed that a national planning framework for wind power should be established, and set a production target of 10TWh per year from offshore wind by 2020. It does not set a target for marine energy development.[132]

E.15.6 Support mechanisms for development of technologies

The Swedish Energy Agency is responsible for Sweden's national energy research programme, financing energy research, technological development and demonstration activities. Under the thematic area of 'the Power System', it is supporting research into offshore wind, wave and marine current energy technologies.[133]

E.15.7 Deployment

In 2009, Sweden installed ten offshore turbines, totalling 30MW at one wind farm. This brought the installed total by the end of 2009 to 75 turbines, with a total capacity of 163.65MW across 5 separate wind farms.[134]

A wave energy converter concept, Seabased, is being tested offshore outside Lysekil, a village on the west coast. The prototype has been developed at Uppsala University and the test unit is in the order of 10kW to 20kW. A very slow speed experimental permanent-magnet generator (5kW) to harness marine current energy has been developed but has not yet been tested in a marine environment.[133]

E.16 US

E.16.1 Offshore wind resource

In 2010, the National Renewable Energy Laboratory produced a report on the opportunities and barriers for offshore wind energy in the US. It estimates that the gross offshore wind resource over all water depths, in regions with annual average wind speeds greater than 8m/s, is 2957GW. When average winds of 7m/s are included this figure increases to 4150GW. Figures E.20 and E.21 indicate the distribution of the US wind resource.

The wind energy resource has been assessed by most states individually. Resource maps for each state are available at the Department of Energy, produced by the National Renewable Energy Laboratory.

E.16.2 Wave resource

Research by the Electric Power Research Institute suggests an available resource of 2100TWh/year at US coasts.[135] Under the assumption of only 15 per cent being utilized, an offshore wave energy resource of 250 to 260TWh/year is calculated. Figure E.22 indicates the distribution of wave energy along the shores of the US.

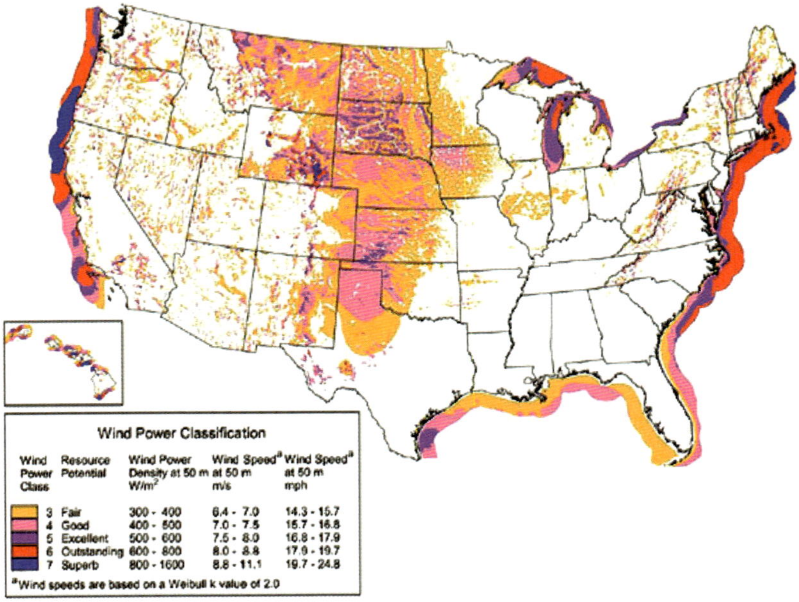

Figure E.20 US land-based and offshore wind resource estimates at 50m height (wind classes 3–7)
Source: NERL (2010)

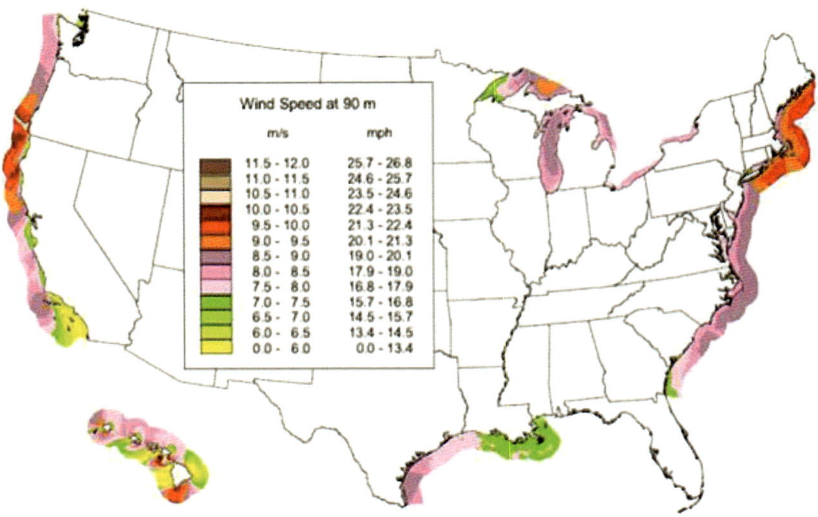

Figure E.21 US offshore wind speed estimates at 90m height
Source: This map was created by the National Renewable Energy Laboratory for the Department of Energy, 2010

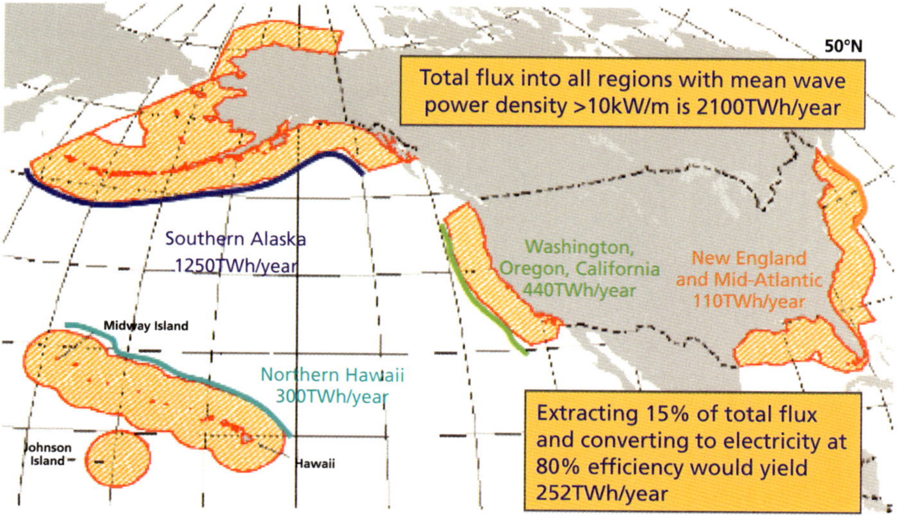

Figure E.22 US wave energy resources
Source: Overview of US Ocean Wave and Current Energy:Resource, Technology,Environmental and Business Issues and Barriers, Sept 2007, p9. Roger Bedard, Ocean Energy Leader EPRI

E.16.3 Tidal resource

In the US, various tidal sites have been assessed regarding their potential energy output. Figure E.23 gives an overview of favourable tidal sites according to the Electric Power Research Institute (EPRI).[136] All of these schemes combined would represent a potential tidal energy resource of about 13TWh/year. The Georgia Tech Research Corporation (GTRC) is currently conducting a more detailed tidal resource assessment.[137]

E.16.4 Processes for permitting, licensing and allocating seabed rights

For wave energy, the relevant federal jurisdictions are:

- plant > 3 miles offshore = Minerals Management Service (MMS)/Federal Energy Regulatory Commission (FERC);
- plant < 3 miles offshore = FERC;
- plant > 3 miles offshore and in a marine sanctuary = National Oceanic and Atmospheric Administration (NOAA)/FERC.

For tidal energy, the relevant federal jurisdictions are:

- commercial plant > 3 miles offshore = FERC;
- pilot plant < 3 miles offshore with no grid connection = United States Army Corps of Engineers (USACE)/FERC.

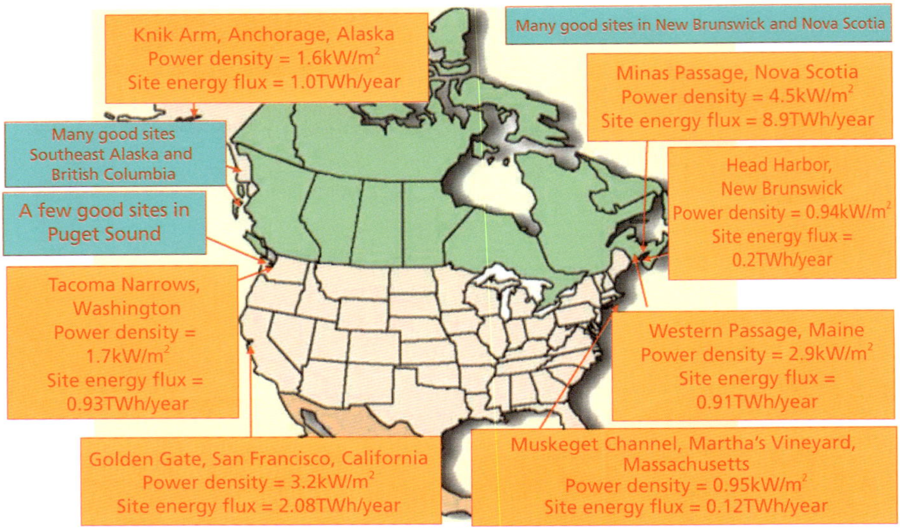

Figure E.23 US tidal energy resources

Source: Overview of US Ocean Wave and Current Energy:Resource, Technology,Environmental and Business Issues and Barriers, Sept 2007, p4. Roger Bedard, Ocean Energy Leader EPRI

Each state also has its own jurisdiction for offshore developments.[138] For most states, this extends to 3 nautical miles from the shoreline, in accordance with the Submerged Lands Act (2002).[139]

There appears to be some debate over which federal department issues permits for the Outer Continental Shelf (OCS), which could affect project development. The Department of the Interior's Minerals Management Service (MMS) indicates that the Energy Policy Act of 2005 gave it authority as the lead agency for projects proposed on the OCS. In November 2007, MMS announced the interim policy for regulating renewable energy on the OCS. It did not allow offshore wind projects to proceed, but established a leasing process for meteorological or marine data collection for five years. In April 2010, MMS released the first request for information (RFI) for commercial leasing for wind power on the OCS, for Delaware, under the new renewable energy framework.[140] The Federal Energy Regulatory Commission (FERC) also states that it has issued preliminary permits for ocean energy projects on the OCS and in the near-shore ocean environment. FERC is also developing a specific licensing process.[141] In 2010 a National Ocean Council was created in order to strengthen ocean governance and coordination across the federal government.[140]

E.16.5 Support mechanisms for project deployment

The US production tax credit (PTC) was extended for wind power through 2012, and for biomass, geothermal, hydropower and offshore power through 2013.[142]

The Emergency Economic Stabilization Act of 2008 expanded the list of eligible technologies to include wave, tidal, current and ocean thermal energy. It applies to systems over 150kW that are in-service by the end of 2013. They receive a credit of US$1.1 cents/kWh (€0.8 cents/kWh). Wind energy installed by the end of 2012 receives a credit of US$2.1 cents/kWh (€1.5 cents/kWh).

President Barack Obama has urged a national US standard requiring 25 per cent renewable power by 2025. Under the Waxman-Markey cap-and-trade bill that narrowly passed the US House in June 2010 and currently awaits reconciliation with its Senate equivalent bill, utilities would be required to meet a goal of 20 per cent renewable energy by 2020. Currently, about 30 states have renewable portfolio standards in place.

The US Department of the Environment (DOE) has a target to deploy 37GW of marine and hydrokinetic (MHK) capacity by 2050[143] and 54GW of offshore wind capacity by 2030.[140]

E.16.6 Support mechanisms for development of technologies

The US accounted for about 53 per cent of reported RD&D funding provided by IEA governments from 1974 to 2003.[144] The DOE manages a Waterpower RD&D programme which began in 2008 at US$10 million, increased to $40 million in 2009 and to $50 million (€37 million) in 2010.[145] The programme is split into two components: marine and hydrokinetics (MHK) and conventional hydropower. The MHK component is allocated US$35 million (i.e. 70 per cent) of the programme funding in 2010. It supports various offshore energy technology projects in wave energy, current energy (including ocean, tidal and river currents) and ocean thermal energy conversion (OTEC).

Government grants for developing new energy technologies are also available for small businesses, up to the value of US$100,000 (€73,000) and with the option to subsequently compete for larger awards. The DOE is currently conducting a national MHK Testing Facilities Needs Assessment in order to identify key gaps in the testing infrastructure in the US. The NREL will issue the report in early 2011.[143]

E.16.7 Other support initiatives

US$50 million (€37 million) annually from 2008 to 2012 has been set aside for the creation of at least one national ocean energy research centre.[146]

E.16.8 Deployment

Thirteen offshore wind projects have advanced significantly along the US permitting process, together totalling around 2.5GW of capacity. These are shown in Figure E.24.[147]

Ocean Power Technologies (OPT) has deployed small-scale prototypes of the PowerBuoy in the waters off Hawaii and New Jersey. Its next generation prototype (150kW) will be deployed for testing in Oregon.[148]

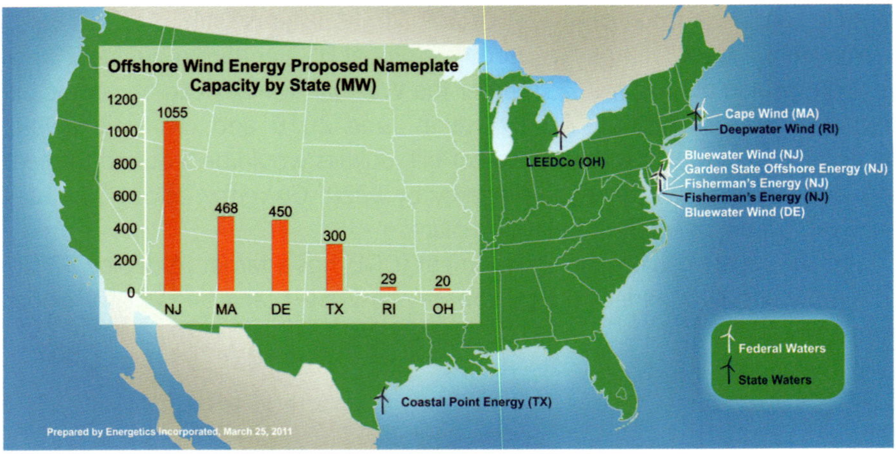

Figure E.24 Proposed US offshore wind projects and capacity, showing projects with significant progress

Source: This map was created by the National Renewable Energy Laboratory for the Department of Energy, 2010

Locations where preliminary wave power plant permits have been issued are shown in Figure E.25.

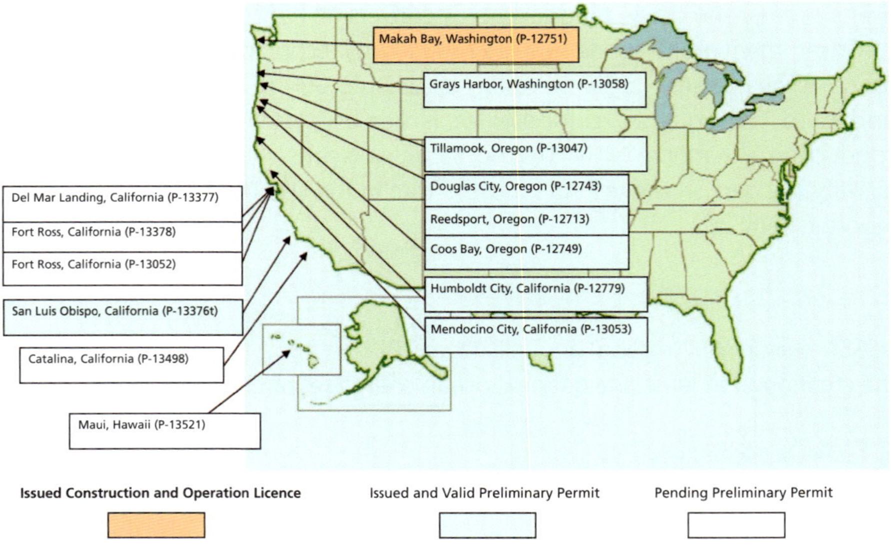

Figure E.25 US wave power plant preliminary permit locations

Source: Wave and Hyrokinetics Interest Group 1st meeting of 2009/2010 Year: With a Focus on Wave Energy, 2009, p8. Roger Bedard, Ocean Energy Leader EPRI

E.17 China

E.17.1 Offshore wind resource

A publication by the China Wind Power Centre considers offshore areas with shallow waters (2m to 15m) at the east coast. The preliminary estimate amounts to a potential of about 750GW.[149] Figure E.26 gives an overview of the wind speeds experienced in eastern China. Coastal regions are exposed to wind speeds of 7.5m/s and above.

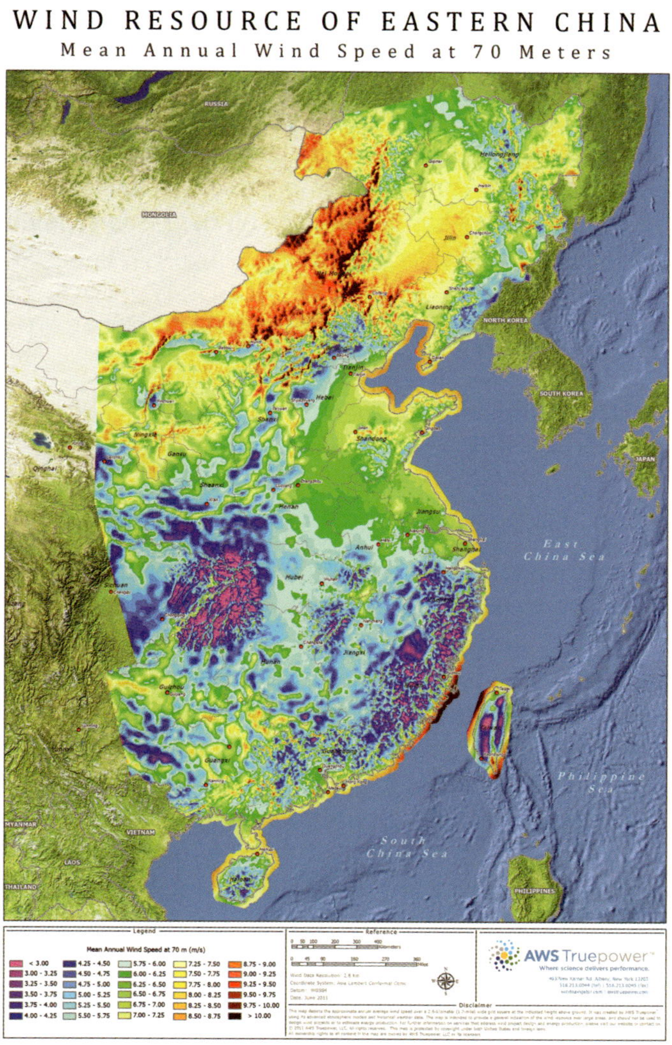

Figure E.26 Wind Resource Map of Eastern China at 70m heights and 3km resolution (mean annual wind speed)

Source: AWS Truepower, 2006

The China Meteorological Association (CMA) and Sgurr Energy recently completed an assessment of offshore wind energy potential in China, assessing the resource along a 10,000km stretch of coastline from Fujian to Shandong. The results are to be published in early 2011.[150]

E.17.2 Wave resource

China has more than 18,000km of coastline which is exposed to a moderate wave climate, with a wave energy flux between 10 and 20kW/m.[151] Studies assessing the actual power generation potential have not been identified.

E.17.3 Tidal resource

While no tidal resource assessment can be referenced, China has a good tidal resource.[152] In particular, the south-eastern coastal areas of Zhejiang, Fujian and Guangdong provinces are considered to have substantial potential for tidal energy.[153] Since 1956, China has been exploiting this resource at a number of sites with small-scale schemes.

E.17.4 Processes for permitting, licensing and allocating seabed rights

Publicly available information on permitting, licensing and seabed rights processes in China was not readily identified.

E.17.5 Support mechanisms for project deployment

The National Development and Reform Commission (NDRC) issued its *Medium- and Long-Term Development Plan for Renewable Energy* in September 2007. The plan establishes targets for the development of various sources of renewable energy up to 2020, calling for the percentage of renewable energy to rise to 10 per cent of total energy consumption by 2010 and 15 per cent by 2020.[154] It states that China will actively promote the development and utilization of ocean energies and sets a total capacity target for tidal power generation of 100MW by 2020. Development of one to two offshore wind farms of 100MW are targeted for 2010, and a total installed capacity of 1GW by 2020. It also introduced a mandatory market share (MMS) of non-hydro renewable generation.[155]

China established a Renewable Energy Law in 2005. This includes provisions, for example, for energy authorities to undertake resource surveys and set out renewable development plans, develop guidance and technical support, and provide economic incentives. While the law does not specify tariffs and technologies, it does refer to ocean energy as one of the relevant technologies.[156] Some amendments to the law were passed in December 2009, introducing renewable obligations for grid companies, putting the MMS policy into legislation. Specific targets are still to be set by the relevant Chinese government agencies.[157]

E.17.6 Support mechanisms for development of technologies

The Renewable Energy Development Fund has been set up through the 2009 Renewable Energy Law amendments. It will collate the surcharge paid by end-users for the additional cost of renewable energy (set at 0.004 Chinese yuan renminbi/kWh (~ €43 cents/MWh) in November 2009). This will create an estimated fund of US$1 billion in 2010 that will compensate grid companies for the extra costs incurred and also provide investment for renewable energy projects and R&D.[157]

The NDRC has put offshore wind development as one of the major R&D priorities in its *Recommendations for Improving the Effectiveness of Renewable Energy in China* guidelines.[158]

E.17.7 Deployment

The first offshore wind turbine in China was installed in 2007, located in Liaodong Bay in the north-east Bohai Sea and with a capacity of 1.5MW. It was built by the China National Offshore Oil Corp (CNOOC). Construction of the first offshore wind farm in China, consisting of 34 3MW turbines, started in 2009 and is now operational. It generated electricity for the 2010 Shanghai Expo. Three 3MW turbines are also expected to be installed in Rudong Offshore Wind Farm in Jiangsu Province. Other provinces that have plans for offshore wind farms are Shandong, Zhejiang and Fujian provinces.[159]

China does not have non-barrage tidal plants.[160]

E.18 Taiwan

E.18.1 Offshore wind resource

With average wind speeds at the coasts of between 6m/s and 7m/s, Taiwan has a good wind energy resource.[161] Similar to Japan, however, offshore development faces the challenge of deep waters surrounding the country, with water depth increasing rapidly and close to shore to many 100m (deepest point reaching 5.5km). Studies assessing the actual power generation potential have not been identified.

E.18.2 Wave resource

Taiwan's coasts are exposed to a good wave regime with a wave energy flux of 15 to 30kW/m.[162]

E.18.3 Tidal resource

Taiwan has a considerable tidal energy resource. Moreover, Taiwan is investigating the potential of ocean currents, with recent estimates amounting to an estimated potential of 60GW.[163] The strongest tidal currents develop between the Island of Taiwan and China's mainland, located at the west coast of Taiwan. Currents surrounding Taiwan are presented in Figure E.27. On the west coast, the Kuroshio Current passes the island close to shore.

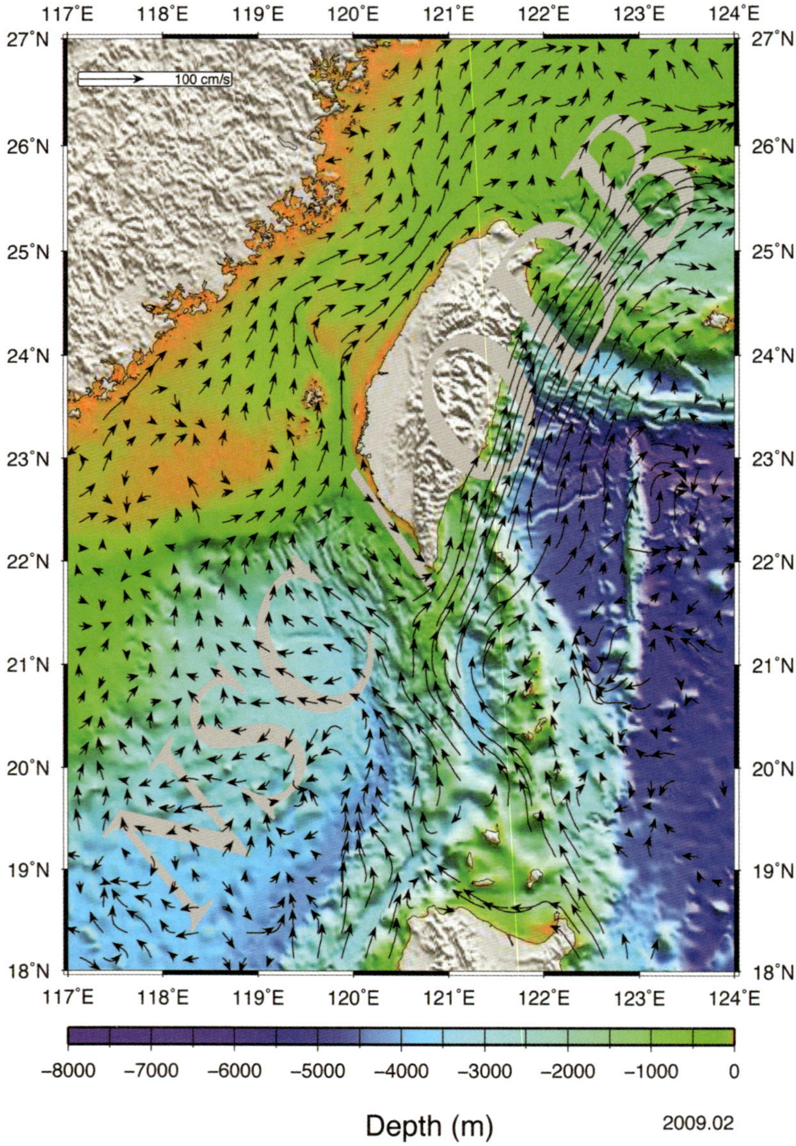

Figure E.27 Taiwan current velocity vectors

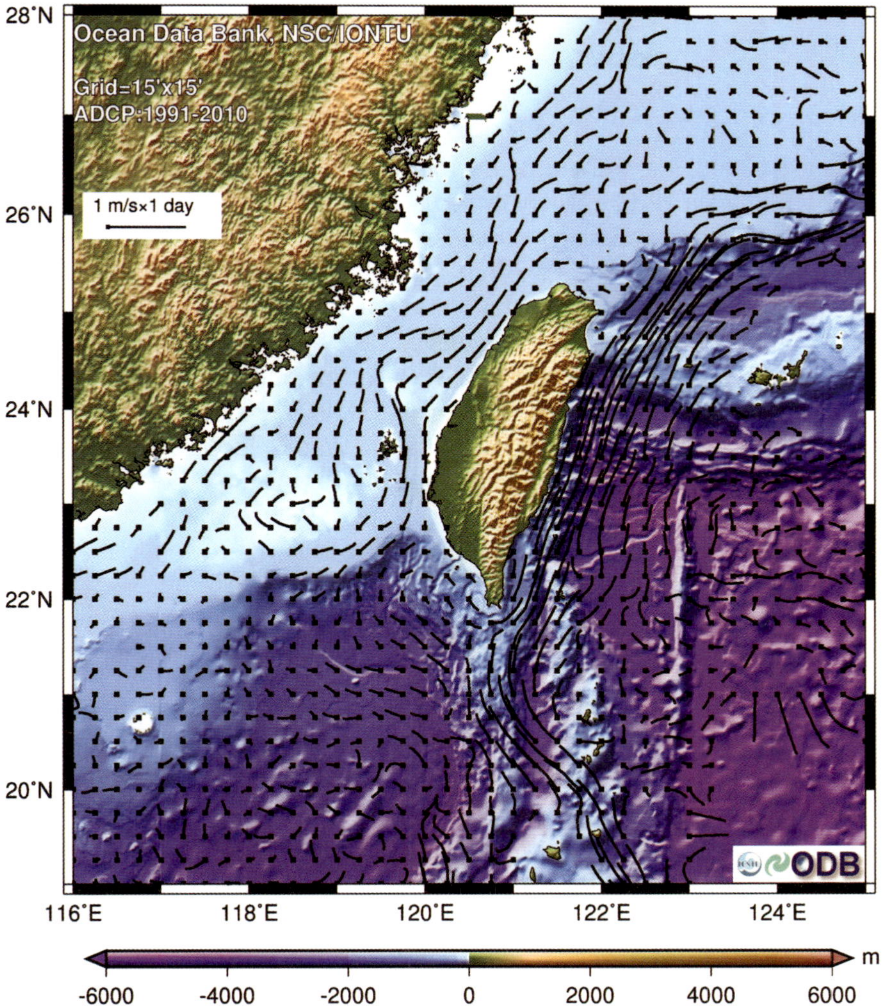

Figure E.27 continued

Source: Ocean Data Bank of National Science Council of Taiwan/Institute of Oceanography, National Taiwan University,2010

E.18.4 Processes for permitting, licensing and allocating seabed rights

Publicly available information on permitting, licensing and seabed rights processes in Taiwan was not readily identified.

E.18.5 Support mechanisms for project deployment

In June 2009, Taiwan's parliament approved a renewable energy bill that calls to add about 6.5GW to 10GW of renewable energy capacity by 2030. Currently, there is minimal capacity deployed.[164] Technology-specific FIT levels are due to be finalized by March 2010.

E.18.6 Support mechanisms for development of technologies

Starting in 2008, the Bureau of Energy, Ministry of Economy Affairs (BOE, MOEA) funded a US$6 million three-year project to study marine energy, including ocean thermal energy, ocean wave energy and ocean current energy.[164] The Industrial Technology Research Institute (ITRI), a government-funded non-profit research institute and the Orkney-based marine renewable company, Aquatera Ltd, in Scotland, have commenced a joint study to assess a potential test site for wave energy converter (WEC) technology. The forecasted wave energy development is expected to be about 75MW by 2025, should the testing be successfully accomplished.[165] Joint with the Water Resource Agency of the MOEA, Taiwan Power Company, the state utility company, is studying a hybrid ocean thermal energy conversion (OTEC) system on the south-east coast. The National Science Council under the Executive Yuan is currently drafting a research plan to explore harnessing the Kuroshio Current energy.[164]

E.18.7 Deployment

Taiwan is moving closer to having its first offshore wind farm, the 600MW Changhua Offshore Wind Farm, which will be located in the Taiwan Strait. The companies involved are the Taiwan Generation Corporation (Taipower) and the European SeaEnergy Renewables.[166]

There is currently no ocean energy in Taiwan; but four potential sites for deep-water ocean current turbine deployment have been proposed in the Kuroshio Current area by local researchers.[164]

E.19 Portugal

E.19.1 Offshore wind resource

Portugal has potential for offshore wind. The continental shelf ends 250km off the coast with the potential of 10GW or greater. Within this 250km barrier there is a limited potential of 3.5GW.[167] Assuming a 35 per cent capacity factor, this would equate to a potential of approximately 11TWh.

E.19.2 Wave resource

There are favourable wave power levels of 30 to 40kW/m within Portuguese waters, with a total of around 21GW of exploitable power.[168] Assuming a 35 per cent capacity factor, this would equate to a potential of approximately 64TWh.

E.19.3 Tidal resource

Portugal has a very low tidal resource; tidal speeds range between 0 to 1 knots (0 to 0.5m/s). Tidal devices operate at speeds above this, making the tidal resource off the Portuguese coast unsuitable.

E.19.4 Government bodies for permitting, licensing and allocating seabed rights

The wave energy pilot zone where wave devices are tested is managed by Redes Energéticas Nacionais (REN). A team with representatives from REN, Sociedade Gestora de Participações Sociais (SGPS) and Sociedade Anónima (SA) work together to make up a company that is responsible for the pilot zone. It is expected that this arrangement will allow a one-stop shop for licences, permits and issuing allocations.[169]

E.19.5 Support mechanisms for project deployment

There are various support mechanisms for marine project deployment, including FITs, subsidies and tax incentives. The FIT scheme was modified in 2005 and introduced a new tariff for emerging technology which includes wave energy. The FITs for offshore wind and wave are €7.4 cents/kWh and €7.6 to €26 cents/kWh, respectively, with 15-year validity. The tariff for wave energy varies with the maturity of the technology, the stage of development and the amount of involvement of Portuguese organizations. There is no tariff for tidal energy.[170]

Further investment subsidies were set up under Decree Portaria 1463/2007 with the following being available for renewable energy. The SIEST programme started in 2006 and provides 25 per cent of total investment for all residential and commercial technologies within the Azores autonomous region. A separate scheme that started in 2007 (due to end in 2013) gives a 35 per cent subsidy of the total invested, with a maximum of €250,000 per project. Companies can also apply for a financial incentive under another programme that provides them with interest-free loans up to €750,000 for five years and two years for capital within the same timeframe.[171]

E.19.6 Support mechanisms for development of technologies

The main supporting institution for Portugal's marine energy is the Instituto Superior Técnico (IST) research facility at the University of Lisbon. This facility, the National Institute of Engineering and Industrial Technologies (INETI) and the Portuguese Ministry of Economy have worked together on various research projects related to marine energy with their main focus being the OWC wave energy converters.[172]

Portuguese research and development funding can come from several places: the Portuguese Ministry of Science and Technology, the Ministry of Economy and the European Commission.[173]

The Portuguese government has spent an estimated €6.8 million on ocean energy since 1980, averaging approximately €0.22 million per year.[174]

Another funding initiative is DEMTEC, which is a programme promoting innovation and supporting research and development of new technologies, especially new pilot research projects. DEMTEC offers grants to developers piloting schemes that fit into this category. In 2006 the budget was about €5 million. The DEMTEC programme was initially set up to fund the renovation of the Pico OWC device.[172, 175]

Portugal also has the Wave Energy Centre (WaveEC). This non-profit centre provides information and supports those companies which are interested in using Portugal for wave energy testing and demonstrations. WaveEC also organizes and takes part in wave energy projects and provides training.[176]

A wave energy pilot zone has been set up to allow for pre-commercial and commercial wave devices producing electricity to be tested. This pilot zone is off the central west coast, 130km north of Lisbon. The zone benefits from a high wave energy capacity of 30kW/m.[177]

E.19.7 Other support initiatives

Portugal is one of the three original members of the IEA Implementing Agreement on Ocean Energy. The WaveEC in Portugal organizes the common fund for the agreement.[178]

Portugal is also one of the members of the Sustainable Economically Efficient Wave Energy Converter (SEEWEC) project. This project promotes a point absorber device to harness wave power off the European coast. The main aim of this project is to assist in developing this converter using experience from the project members.[179]

E.19.8 Deployment

The world's first wave farm was deployed in 2008 and consisted of three Pelamis 750kW devices, which were installed 5km from the Portuguese coast. The aim of the project was to increase the number of devices to 25 and produce 21MW of power. However, two months after deployment, the project experienced technical problems which required devices to be brought back to shore. The project subsequently encountered financial difficulties as the project owner went into receivership. Efforts to find a new owner appeared to have failed. Lessons learned from the Aguçadoura project have nevertheless allowed Pelamis Wave Power to improve their design for their second-generation P2 device soon to be deployed in the UK.[180]

Pico Island, part of the Azores group of islands, hosts a 400kW OWC device. The Pico OWC started operation in 1999 and was funded by the European Commission, Electricidade dos Acores (EDA) – the original owner – and Electricidade de Portugal (EDP). Under the coordination of the IST, the device became operational; but due to technical issues the project was interrupted for several years. A recovery project costing €600,000 in public funding and a matched amount of private funding, including money from the EDP, EDA, EFACEC, Consulmar, Irmãos Cavaco, IST and INETI, was set up to improve the device and restart operation. However, due to design limitations, the OWC will not operate at 100 per cent efficiency. This is an on-going project for WaveEC (the new owner) and is used for demonstration purposes.[181]

Notes

1 Fugro Oceanor (2008).
2 Canadian Hydraulics Centre (2006).
3 Canadian Hydraulics Centre (2006).
4 Canadian Hydraulics Centre (2006).
5 Fisheries and Oceans Canada (2010).
6 IEA (2010).
7 Invest in Canada (2010).
8 Pers comm with Natural Resources Canada (2010).
9 Canadian Hydraulics Centre (2006).
10 EWEA (2010a).
11 EPRI (2009).
12 Duwind (2001).
13 Marine Institute, Sustainable Energy Ireland (2002).
14 RAMBØLL (1999).
15 Danish Energy Agency (2010).
16 IEA (2010).
17 EREC (2009).
18 EWEA (2010a).
19 Danish Energy Agency (2010).
20 Duwind (2001).
21 Fugro Oceanor (2008).
22 Clément et al (2002).
23 GWEC (2010).
24 DGEC (2010).
25 EREC (2009).
26 GWEC (2010).
27 IEA (2010).
28 ADEME (2009).
29 EurObserv'ER (2010).
30 WEC (2009).
31 BWE (2010).

32 Duwind (2001).
33 German Environment Ministry (2007).
34 German Environment Ministry (2009).
35 German Environment Ministry (2007).
36 EWEA (2010a).
37 German Environment Ministry (2007).
38 EWEA (2010a).
39 Duwind (2001).
40 Clément et al (2002).
41 Marine Institute, Sustainable Energy Ireland (2005).
42 Marine Institute, Sustainable Energy Ireland (2002).
43 SEI (2005).
44 SEI (2005).
45 Department of the Environment, Heritage and Local Government (2010).
46 SEAI (2010).
47 EREC (2009).
48 EWEA (2010a).
49 Duwind (2001).
50 Fugro Oceanor (2008).
51 Watson, Farley and Williams (2008).
52 EREC (2009).
53 Blue H Group (2010).
54 EurObserv'ER (2010).
55 Takagi et al (2006).
56 WEC (2007).
57 Fugro Oceanor (2008).
58 METI (2010).
59 NEDO (2009).
60 JAMSTEC (2010).
61 NEDO (2008).
62 IEA–OES (2009a).
63 Duwind (2001).
64 Fugro Oceanor (2008).
65 IEA Wind (2009).
66 EREC (2009).
67 EWEA (2010a).
68 EurObserv'ER (2010).
69 NEV (2008).
70 Kjeller Vindteknikk (2009).
71 Enova (2007).
72 Fugro Oceanor (2008).
73 Marine Institute, Sustainable Energy Ireland (2002).
74 Clément et al (2002).
75 Enova (2007).
76 Nortrade.com (2009).
77 Sway (2010).

78 NORCOWE (2010).
79 NOWITECH (2010).
80 EWEA (2010a).
81 IEA–OES (2009a).
82 Sway (2010).
83 Nortrade.com (2009).
84 Duwind (2001).
85 BERR (2008).
86 Thorpe (1992).
87 Carbon Trust (2006).
88 Arup Energy/The Carbon Trust (2005).
89 Black & Veatch/The Carbon Trust (2005).
90 Couch (2010); see http://www.all-energy.co.uk/userfiles/file/andy-baldock-180511.pdf for a summary.
91 Black & Veatch/The Carbon Trust (2005).
92 GWEC (2010).
93 The Crown Estate (2010b).
94 IEA–OES (2009a).
95 BWEA/Entec (2009).
96 HM Government (2009).
97 UKERC (2008).
98 BWEA/Entec (2009).
99 DECC (2010b).
100 NFPA (2010).
101 BWEA/Entec (2009).
102 BWEA/Entec (2009).
103 Scottish Government (2010c).
104 The Crown Estate (2010b).
105 EWEA (2010a).
106 WEC (2009).
107 BWEA/Entec (2009).
108 The Crown Estate (2010a).
109 Van Hulle et al (2004).
110 Duwind (2001).
111 Fugro Oceanor (2008).
112 EREC (2009).
113 EWEA (2010a).
114 Duwind (2001).
115 Fugro Oceanor (2008).
116 EREC (2009).
117 German Environment Ministry (2010).
118 EWEA (2010a).
119 Duwind (2001).
120 Fugro Oceanor (2008).
121 Carballo et al (2008).
122 Spanish Ministry of Environment and Rural and Marine Areas (2009).

123 German Environment Ministry (2010).
124 GWEC (2010).
125 Ente Vasco de la Energia (2009).
126 International Water Power & Dam Construction (2010).
127 BWEA/Entec (2009).
128 Ente Vasco de la Energia (2009).
129 Duwind (2001).
130 Fugro Oceanor (2008).
131 Swedish Energy Agency (2009b).
132 Ministry of the Environment and Ministry of Enterprise, Energy and Communications (2009).
133 Swedish Energy Agency (2009a).
134 EWEA (2010a).
135 EPRI (2009).
136 EPRI (2007).
137 DOE (2009).
138 EPRI (2009).
139 NREL (2010).
140 NREL (2010).
141 Lane (2008).
142 REN21 (2009).
143 Reed (2010); for an update, see http://www.cleanenergystates.org/resource-library/resource/update-on-the-marine-and-hydrokinetic-energy-testing-needs-and-capabilities-assessement0.
144 IEA (2006).
145 EPRI (2009).
146 Lane (2008).
147 NREL (2010).
148 BWEA/Entec (2009).
149 Pengfei (2008).
150 CMA and Sgurr Energy (forthcoming).
151 Fugro Oceanor (2008).
152 Lyard et al (2006).
153 WEC (2007).
154 IEA (2010).
155 NDRC (2007).
156 NDRC (2005).
157 Schuman (2010).
158 REN 21 (2009).
159 EWEA (2010a).
160 IEA–OES (2009a).
161 Risø National Laboratory for Sustainable Energy (undated).
162 Fugro Oceanor (2008).
163 Barr (2009), citing Professor Chen Falin, Institute of Applied Mechanics, National Taiwan University.
164 Chuang et al. (2009).

165 Aquatera Ltd (2008).
166 EWEA (2010a).
167 *Windpower Monthly* (2008).
168 Wave Energy Centre (2004).
169 IEA–OES (2009a).
170 EREC (2009).
171 EREC (2009).
172 Simoes et al. (2007).
173 IEA (2010).
174 IEA (2009b).
175 Wave Energy Centre (2010b).
176 Wave Energy Centre (2010a).
177 IEA–OES (2009b).
178 IEA–OES (2009a).
179 SEEWEC (2010).
180 CleanTech (2010).
181 Wave Energy Centre (2010b).

Index